Remediation Strategies for Soil and Water

Remediation Strategies for Soil and Water

Guest Editor

Xiaoqiang Cui

MDPI

Basel • Beijing • Wuhan • Barcelona • Belgrade • Novi Sad • Cluj • Manchester

Guest Editor
Xiaoqiang Cui
School of Environmental
Science and Engineering
Tianjin University
Tianjin
China

Editorial Office
MDPI AG
Grosspeteranlage 5
4052 Basel, Switzerland

This is a reprint of the Special Issue, published open access by the journal *Processes* (ISSN 2227-9717), freely accessible at: https://www.mdpi.com/journal/processes/special_issues/BOTY607CL0.

For citation purposes, cite each article independently as indicated on the article page online and as indicated below:

Lastname, A.A.; Lastname, B.B. Article Title. *Journal Name* **Year**, *Volume Number*, Page Range.

ISBN 978-3-7258-4783-9 (Hbk)
ISBN 978-3-7258-4784-6 (PDF)
https://doi.org/10.3390/books978-3-7258-4784-6

Contents

About the Editor

Xiaoqiang Cui

Xiaoqiang Cui is an Associate Professor at the School of Environmental Science and Engineering, Tianjin University. Xiaoqiang received a B.Sc. degree in Environmental Engineering from Southwest University in 2013 and a Ph.D. degree in Environmental Remediation and Resource Regeneration from Zhejiang University in 2018. His research interests mainly include the green remediation of contaminated soil and water, the harmless treatment and value-added utilization of biomass wastes, and the syntheses of functional carbonaceous materials. He serves on the editorial board/young editorial board of *Processes, Carbon Research,* and the *Chinese Journal of Soil Science.* Additionally, he has conducted 13 scientific research projects and published more than 70 papers in SCI journals.

processes

MDPI

Editorial

Remediation Strategies for Soil and Water

Junxia Wang and Xiaoqiang Cui *

Tianjin Key Lab of Biomass Waste Utilization, School of Environmental Science and Engineering, Tianjin University, Tianjin 300072, China; wangjunxia@tju.edu.cn
* Correspondence: cuixiaoqiang@tju.edu.cn

With the rapid development of industry worldwide, soil and water pollution has increased in recent decades [1,2]. Pollutants such as heavy metals, polycyclic aromatic hydrocarbons, antibiotics, and microplastics in soil and water cause severe environmental problems [3–6], posing serious risks to human health and the ecological system. Approximately 4 billion people, representing about half of the world's population, suffer from severe physical water scarcity for at least part of the year [7]. Hence, several techniques have been developed for the remediation of contaminated soil and water, including adsorption, membrane separation, advanced oxidation, ion exchange, electrochemical treatment, microbial degradation, and phytoremediation [8–15]. Considering the ongoing issue of pollution and emerging pollutants, remediation strategies for soil and water should be further optimized to focus on developing functional materials, improving remediation efficiency, and enhancing integrated remediation processes. This Special Issue entitled "Remediation Strategies for Soil and Water" features six research articles and three review papers, each addressing critical aspects of remediation processes for soil and water. The contributions are briefly described below.

Liu et al. [16] demonstrated the application of the volume-of-fluid (VOF) technique in the prediction of leakage and diffusion of underwater oil spill, providing potential remediation strategies. They determined the impacts of various parameters on leakage and diffusion characteristics by adjusting ocean current velocity, spilled oil velocity, and the density and viscosity of crude oil. Notably, a user-defined function was proposed and used for computational fluid dynamics modeling, and they obtained a critical formula to determine the lateral migration distance of spilled oil under various conditions. Their results can provide a theoretical basis for developing an emergency treatment scheme for offshore oil spills and improve the remediation efficiency of oil-contaminated water.

Yang et al. [17] focused on the effective removal of ammonium from water using ball-milled biochar modified with NaOH. They aimed to explore the feasibility of using modified biochars to achieve the enhanced remediation of ammonium-polluted water. The results indicate that the ball-milling treatment and NaOH modification greatly improved the ammonium sorption capacity of biochar, and the modified biochars showed enhanced sorption performance in a wide pH range (3–7). The surface complexation with functional groups and cation exchange were considered the main sorption mechanisms. The authors concluded that biochar modified via NaOH/ball-milling can be employed as a cost-effective sorbent for the remediation of ammonium-polluted water.

Meng et al. [18] proposed a promising scheme for the sustainable remediation of nitrate-contaminated groundwater through permeable reactive barrier (PRB) remediation systems. They fabricated nano zero-valent iron (nZVI) particles encapsulated within an acrylonitrile–butadiene–styrene (ABS) matrix, which efficiently solved the problem of aggregation and reactivity decrease through regulating the distribution of nZVI particles and controlling the interaction between nZVI particles and the polluted water. The results demonstrate that encapsulation suppressed the aggregation of nZVI, achieving sustainable denitrification performance during wastewater remediation.

Luo et al.'s research [19] delves into the impacts of fertilization on nitrogen and phosphorus pollution in agricultural water bodies. They built a crop–soil runoff water system to determine the variation in crop growth and the pollutant contents in water bodies under different fertilization rates. The authors emphasized that crop growth intercepted the migration of nitrogen and phosphorus, which was further affected by different growth stages. Considering the combined effects of fertilization on soil fertility, crop yield, and environmental risk, they proposed an optimized fertilization rate for an eggplant planting system, achieving the best interception effect on nitrogen and phosphorus in the water system.

Wang et al. [20] provide an overview of a hot topic, reducing greenhouse gas (GHG) emissions from farmland soil, especially focusing on the microbial-mediated process. They systematically illuminated the main generation pathways of GHGs (i.e., CO_2, CH_4, and N_2O) in soil and analyzed the impacts of various environmental factors, including soil temperature, organic matter, moisture, and pH, on soil GHG emissions. Additionally, they determined the microbial mechanisms involved in the soil GHG emission process under various soil remediation modes (e.g., the biochar addition, organic fertilizer addition, straw return, and microalgal biofertilizer application modes). They demonstrated the critical role of microalgae biofertilizer in reducing GHG emissions in farmland soil, providing a theoretical basis for achieving carbon neutrality in agriculture production.

Wang et al. [21] systematically reviewed hydrothermal treatment for food waste to address its potential environmental risk. The authors presented an in-depth analysis of the effects of the mechanisms of hydrothermal conditions (e.g., hydrothermal temperature, time, feedstock properties, and catalysts) on the generation of various products, focusing on the potential application of the value-added products derived from the hydrothermal treatment of food waste. Hydrochar, bio-oil, and gaseous products derived from this process showed good potential in environmental remediation, soil improvement, energy production, and new material synthesis. Finally, the authors proposed the potential challenges and development directions for the hydrothermal treatment of food waste.

Liu et al. [22] conducted a study on landfill gas–water joint regulation technology. Low oxygen utilization rates resulted in high energy consumption and operating costs, further restraining the application of in situ aeration technology to stabilize municipal solid waste. The authors aimed to improve oxygen utilization efficiency through observing the dynamic respiratory index and the removal of pollutants. Hence, they constructed three continuous reactors and proposed targeted aeration and re-circulation schemes for various landfill ages. The results imply that a well-designed reactor (aerobic, semi-aerobic, and anaerobic systems) can effectively degrade the organic components of municipal solid waste with various landfill ages, and gas–water joint technology is beneficial to activating microorganisms.

Cao et al. [23] provide a comprehensive overview of oxygenated and nitrated polycyclic aromatic hydrocarbons (PAHs) in soil, including the sources, quantification, incidence, toxicity, and transport of these PAHs in soil. Considering the high toxicity and distinct environmental effects of oxygenated and nitrated PAHs, their behaviors were systematically summarized. The authors discussed the toxicity, human health risk, and transformation of oxygenated and nitrated PAHs in soil, providing a valuable basis for the efficient remediation of PAH-contaminated soil.

The study by Ortiz-Soto et al. [24] focused on the electrokinetic remediation of heavy-metal-contaminated mining area soil. The electrokinetic remediation (EKR) technique was employed to assess the migration of heavy metal contaminant from fresh mine tailings.

Hence, a pilot EKR cell was designed to study the recovery potential of cobalt, nickel, and copper species, and the impacts of electric field intensity, pH, and their interaction on heavy metal recovery performance were determined. The authors found the optimized electric field and pretreatment solution concentration for the enhanced migration of typical heavy metals, promoting the recovery of various species and the remediation of mining area soil.

This Special Issue aims to elucidate the recent advancements and emerging challenges in the remediation of soil and water. We hope that these valuable contributions from outstanding scientists will inspire readers to gain new ideas for future research and promote the development of environmental remediation.

Author Contributions: Conceptualization, and writing—original draft preparation, J.W.; conceptualization, writing—review and editing, and supervision, X.C. All authors have read and agreed to the published version of the manuscript.

Conflicts of Interest: There are no conflicts of interest to declare.

References

1. Lu, Y.; Song, S.; Wang, R.; Liu, Z.; Meng, J.; Sweetman, A.J.; Jenkins, A.; Ferrier, R.C.; Li, H.; Luo, W.; et al. Impacts of soil and water pollution on food safety and health risks in China. *Environ. Int.* **2015**, *77*, 5–15. [CrossRef] [PubMed]
2. Schwarzenbach, R.P.; Egli, T.; Hofstetter, T.B.; von Gunten, U.; Wehrli, B. Global Water Pollution and Human Health. *Annu. Rev. Environ. Resour.* **2010**, *35*, 109–136. [CrossRef]
3. Hou, D.Y.; O'Connor, D.; Igalavithana, A.D.; Alessi, D.S.; Luo, J.; Tsang, D.C.W.; Sparks, D.L.; Yamauchi, Y.; Rinklebe, J.; Ok, Y.S. Metal contamination and bioremediation of agricultural soils for food safety and sustainability. *Nat. Rev. Earth Environ.* **2020**, *1*, 366–381. [CrossRef]
4. Al-sareji, O.J.; Grmasha, R.A.; Jakab, M.; Khan, M.S.; Islam, S.; Biswas, T.; Meiczinger, M.; Stenger-Kovacs, C. PAHs in the sediments of the Shatt Al-Hillah River, Babylon, Iraq: Spatial and temporal distribution, pollution sources, human impacts, and treatment strategy. *Process Saf. Environ. Prot.* **2025**, *200*, 107371. [CrossRef]
5. Sarmah, A.K.; Meyer, M.T.; Boxall, A.B.A. A global perspective on the use, sales, exposure pathways, occurrence, fate and effects of veterinary antibiotics (VAs) in the environment. *Chemosphere* **2006**, *65*, 725–759. [CrossRef]
6. Horton, A.A.; Walton, A.; Spurgeon, D.J.; Lahive, E.; Svendsen, C. Microplastics in freshwater and terrestrial environments: Evaluating the current understanding to identify the knowledge gaps and future research priorities. *Sci. Total Environ.* **2017**, *586*, 127–141. [CrossRef]
7. UN-Water. UN World Water Development Report 2025. Available online: https://www.unwater.org/publications/un-world-water-development-report-2025 (accessed on 21 March 2025).
8. Cui, X.; Wang, J.; Wang, X.; Khan, M.B.; Lu, M.; Khan, K.Y.; Song, Y.; He, Z.; Yang, X.; Yan, B.; et al. Biochar from constructed wetland biomass waste: A review of its potential and challenges. *Chemosphere* **2022**, *287*, 132259. [CrossRef]
9. Ighalo, J.O.; Chen, Z.; Ohoro, C.R.; Oniye, M.; Igwegbe, C.A.; Elimhingbovo, I.; Khongthaw, B.; Dulta, K.; Yap, P.-S.; Anastopoulos, I. A review of remediation technologies for uranium-contaminated water. *Chemosphere* **2024**, *352*, 141322. [CrossRef]
10. Wei, K.H.; Ma, J.; Xi, B.D.; Yu, M.D.; Cui, J.; Chen, B.L.; Li, Y.; Gu, Q.B.; He, X.S. Recent progress on in-situ chemical oxidation for the remediation of petroleum contaminated soil and groundwater. *J. Hazard. Mater.* **2022**, *432*, 128738. [CrossRef]
11. Esfahani, E.B.; Dixit, F.; Zeidabadi, F.A.; Johnson, M.R.; Mayilswamy, N.; Kandasubramanian, B.; Mohseni, M. Ion exchange and advanced oxidation/reduction processes for per- and polyfluoroalkyl substances treatment: A mini-review. *Curr. Opin. Chem. Eng.* **2023**, *42*, 100953. [CrossRef]
12. Yang, X.; Liu, L.H.; Tan, W.F., Liu, C.S., Dang, Z., Qiu, G.H. Remediation of heavy metal contaminated soils by organic acid extraction and electrochemical adsorption. *Environ. Pollut.* **2020**, *264*, 114745. [CrossRef]
13. Wang, Y.C.; Li, A.; Cui, C.W. Remediation of heavy metal-contaminated soils by electrokinetic technology: Mechanisms and applicability. *Chemosphere* **2021**, *265*, 129071. [CrossRef] [PubMed]
14. Bose, S.; Kumar, P.S.; Vo, D.V.N.; Rajamohan, N.; Saravanan, R. Microbial degradation of recalcitrant pesticides: A review. *Environ. Chem. Lett.* **2021**, *19*, 3209–3228. [CrossRef]
15. Zhang, J.; Cao, X.; Yao, Z.; Lin, Q.; Yan, B.; Cui, X.; He, Z.; Yang, X.; Wang, C.-H.; Chen, G. Phytoremediation of Cd-contaminated farmland soil via various Sedum alfredii-oilseed rape cropping systems: Efficiency comparison and cost-benefit analysis. *J. Hazard. Mater.* **2021**, *419*, 126489. [CrossRef] [PubMed]
16. Liu, R.; Ding, S.; Ju, G. Numerical Study of Leakage and Diffusion of Underwater Oil Spill by Using Volume-of-Fluid (VOF) Technique and Remediation Strategies for Clean-Up. *Processes* **2022**, *10*, 2338. [CrossRef]

17. Yang, H.; Li, X.; Wang, Y.; Wang, J.; Yang, L.; Ma, Z.; Luo, J.; Cui, X.; Yan, B.; Chen, G. Effective Removal of Ammonium from Aqueous Solution by Ball-Milled Biochar Modified with NaOH. *Processes* **2023**, *11*, 1671. [CrossRef]
18. Meng, F.; Yang, Y.; Li, M.; Zhu, Q.; Qin, B.; Yang, C. Nano Zero-Valent Iron (nZVI) Encapsulated with ABS (nZVI/(ABS + EC)) for Sustainable Denitrification Performance and Anti-Aggregation. *Processes* **2024**, *12*, 697. [CrossRef]
19. Luo, P.; Xin, C.; Zhu, Y.; Liu, Y.; Ling, J.; Wang, T.; Huang, J.; Khu, S.-T. Effect of Rational Fertilizer for Eggplants on Nitrogen and Phosphorus Pollutants in Agricultural Water Bodies. *Processes* **2023**, *11*, 579. [CrossRef]
20. Wang, H.; Zhao, R.; Zhao, D.; Liu, S.; Fu, J.; Zhang, Y.; Dai, N.; Song, D.; Ding, H. Microbial-Mediated Emissions of Greenhouse Gas from Farmland Soils: A Review. *Processes* **2022**, *10*, 2361. [CrossRef]
21. Wang, C.; Wang, Z.; Wang, X.; Li, N.; Tao, J.; Zheng, W.; Yan, B.; Cui, X.; Cheng, Z.; Chen, G. A Review on the Hydrothermal Treatment of Food Waste: Processing and Applications. *Processes* **2022**, *10*, 2439. [CrossRef]
22. Liu, J.; Pan, T.; Zhao, H.; Guo, Y.; Chen, G.; Hou, L. Application of Landfill Gas-Water Joint Regulation Technology in Tianjin Landfill. *Processes* **2023**, *11*, 2382. [CrossRef]
23. Cao, W.; Yuan, J.; Geng, S.; Zou, J.; Dou, J.; Fan, F. Oxygenated and Nitrated Polycyclic Aromatic Hydrocarbons: Sources, Quantification, Incidence, Toxicity, and Fate in Soil-A Review Study. *Processes* **2023**, *11*, 52. [CrossRef]
24. Ortiz-Soto, R.; Leal, D.; Gutierrez, C.; Aracena, A.; Leon, M.; Lazo, A.; Lazo, P.; Ottosen, L.; Hansen, H. Incidence of Electric Field and Sulfuric Acid Concentration in Electrokinetic Remediation of Cobalt, Copper, and Nickel in Fresh Copper Mine Tailings. *Processes* **2023**, *11*, 108. [CrossRef]

processes

Article

Numerical Study of Leakage and Diffusion of Underwater Oil Spill by Using Volume-of-Fluid (VOF) Technique and Remediation Strategies for Clean-Up

Rengguang Liu [1,2], Shidong Ding [1,2] and Guoshuai Ju [3,*]

1 State Key Laboratory of Shale Oil and Gas Enrichment Mechanisms and Effective Development, Beijing 102206, China
2 SINOPEC Research Institute of Petroleum Engineering Co., Ltd., Beijing 102206, China
3 College of Petroleum Engineering, Northeast Petroleum University, Daqing 163318, China
* Correspondence: juguoshuai@nepu.edu.cn

Abstract: An oil spill accident will cause serious harm to marine ecology and the environment. Rapid response and effective prevention methods are required to minimize the damage of oil spill accidents. The critical problems that marine emergency rescue teams face are when the spilled oil reaches the sea surface, the extent of the spilled oil, and how far they are from the drilling platform. However, there is no reliable model to predict the diffusion distance of spilled oil. Accurately predicting the diffusion characteristics of underwater spilled oil can provide timely and accurate information for the treatment of oil spill accidents and guide the correct implementation of emergency treatment. In this paper, the computational fluid dynamics (CFD) method was used to establish a two-phase flow model for the diffusion of a submarine oil spill. The volume-of-fluid (VOF) technique was implemented to track the interface between oil–water phases. The effects of different parameters on leakage and diffusion characteristics were investigated by adjusting spilled oil velocity, ocean current velocity, crude oil density, and crude oil viscosity. The logarithmic velocity profile was adopted for ocean currents to conform to the actual flow near the sea surface. A user-defined function (UDF) was developed and applied for CFD modeling. The focus was on analyzing the diffusion range (rising height H_p and lateral migration distance W_p) from full-field data. The results indicate that the oil spill velocity, ocean current velocity, crude oil density, and crude oil viscosity impact the viscous shear force, the oil spill's inertia force, and the current shear effect. The formula for calculating the lateral migration distance of spilled oil under different working conditions was obtained by fitting. The results of this study can provide a scientific basis for formulating an emergency treatment plan for offshore oil spill accidents and minimizing the harm to marine ecology and the environment.

Keywords: drilling platform; underwater oil spill; polluted water; leakage and diffusion; VOF; remediation strategies

1. Introduction

The exploration and extraction of offshore oil and gas resources have become essential to solving the shortage of limited onshore oil resources globally [1,2]. However, large-scale offshore oil and gas exploration and extraction will inevitably bring the risk of crude oil leakage [3,4]. Many severe underwater oil spill accidents have occurred worldwide in the past decades. For example, the Deepwater Horizon accident in the Gulf of Mexico in 2010 was the most severe offshore oil spill on record. The spill lasted three months, resulting in a large amount of crude oil leaking at sea [5–7].

The spread of oil spills at sea will cause serious harm to the marine ecological environment. Rapid response and effective prevention methods are required to minimize the damage of oil spill accidents [8]. The critical problems that marine emergency rescue teams face are when the spilled oil reaches the sea surface, the extent of the spilled oil, and how far they are from the drilling platform. Therefore, it is of great significance to predict the underwater diffusion characteristics of spilled oil and its migration distance under the action of ocean current for the emergency response of oil spill treatment [9,10].

For underwater oil spills, the oil phase gradually disperses in the process of rising to the water surface under the action of buoyancy, partially dissolves in seawater, and is biodegradable, partially deposited on the seabed and partially surfaced. Underwater oil spills are generally divided into two stages: the diffusion and migration process in seawater and the floating process after oil spills reach the sea surface [11]. Suppose the location of the spilled oil can be deduced according to the marine and meteorological conditions after the spilled oil on the sea surface is discovered. In that case, emergency remedial measures can be taken to plug the leak point in time. After the oil comes out of the water, it can be disposed of by means of emergency recovery, adsorption, combustion, etc. Some of the escaped pollutants migrate with the seawater, and further through volatilization, dissolution, sedimentation, biodegradation, etc. If we can predict the location of the spilled oil before it reaches the sea surface, and then deal with the oil spill at the beginning of its diffusion and drift, this will not only reduce the workload but also reduce the pollution of the marine environment.

As the field test of underwater crude oil leakage and diffusion is difficult, it is almost impossible to capture the entire spilled oil diffusion trajectory through experimental research. With the progress of science and technology, using numerical simulation instead of field tests has attracted the attention of researchers [4,12–14]. Many scholars are committed to the numerical simulation of offshore oil spills to achieve more accurate prediction after oil spills and provide a technical basis for the timely handling of accidents and the impact assessment of accidents.

However, the underwater environment is complex, and various factors will change the leakage and diffusion behavior of spilled oil. In this study, we use the computational fluid dynamics (CFD) method to establish a two-dimensional numerical model of the underwater oil spill. The volume-of-fluid (VOF) method was introduced to track the interface between oil–seawater phases. The simulated results were in agreement with the experimental results of Zhu et al. [15]. The effects of different parameters on leakage and diffusion characteristics were investigated by adjusting spilled oil velocity, ocean current velocity, crude oil density, and crude oil viscosity. The results of this study can provide helpful information for taking emergency response measures to reduce the impact of crude oil spillage.

2. Computational Methodologies

2.1. Governing Equations

Spilled oil and seawater are considered incompressible fluids, and it is assumed that there is no phase transition and slip at the oil–water interface. The volume-of-fluid (VOF) multiphase model is implemented to simulate the diffusion of underwater oil spill [16,17]. The interface tracking between the phases is accomplished by solving the continuity equation for the volume fraction of oil–water in two phases. The VOF method has been continuously improved and widely used for decades since it was proposed. It can naturally keep the mass of the fluid in balance. It can efficiently deal with the large deformation of the free surface and the change of the topology of the free surface. For calculating the fluid volume in a cell, it is only necessary to calculate the fluid volume in the cell adjacent to the cell. Based on this definition, the continuity equations for the oil and water phases can be written as follows.

$$\frac{\partial \alpha_q}{\partial t} + \nabla \cdot (\alpha_q \vec{u}) = 0 \tag{1}$$

where $\vec{u}$ is the velocity vector of the fluid, α is the volume fraction of phase q, and the volume fraction of each phase is calculated based on the following formula:

$$\sum_{q=1}^{n} \alpha_q = 1 \tag{2}$$

The momentum equation is calculated by using a single-velocity field $\vec{u}$ acting on a mixed fluid with a density of $\rho = \alpha_q \cdot \rho_q + \alpha_p \cdot \rho_p$ and a viscosity of $\mu = \alpha_q \cdot \mu_q + \alpha_p \cdot \mu_p$. The momentum equation can be written as follows.

$$\frac{\partial}{\partial t}(\rho\vec{u}) + \nabla \cdot (\rho\vec{u}\vec{u}) = -\nabla p + \nabla \cdot \tau + \rho \cdot \vec{g} + \vec{F} \tag{3}$$

In which, p is the pressure, $\vec{g}$ is the gravity acceleration, τ is the stress tensor, and $\vec{F}$ describes the interaction forces acting on phases such as surface tension.

The realizable *k*-ε turbulence model is used to solve the velocity fluctuation caused by turbulence. Due to the excellent compromise between numerical requirements and stability, the realizable *k*-ε turbulence model has been widely used in industry in recent decades [18–20]. In the realizable *k*-ε model, the turbulent kinetic energy (TKE) *k* and its dissipation rate ε are described by the following two balance equations.

$$\frac{\partial(\rho k)}{\partial t} + \frac{\partial(\rho k \overline{u_j})}{\partial x_j} = \frac{\partial}{\partial x_j}\left[(\mu + \frac{\mu_t}{\sigma_k})\frac{\partial k}{\partial x_j}\right] + G_k - \rho\varepsilon \tag{4}$$

$$\frac{\partial(\rho\varepsilon)}{\partial t} + \frac{\partial(\rho\varepsilon\overline{u_j})}{\partial x_j} = \frac{\partial}{\partial x_j}\left[(\mu + \frac{\mu_t}{\sigma_\varepsilon})\frac{\partial \varepsilon}{\partial x_j}\right] + \rho C_1 S\varepsilon - \rho C_2 \frac{\varepsilon^2}{k + \sqrt{\frac{\mu}{\rho}\varepsilon}} \tag{5}$$

where σ_k and σ_ε are turbulent Prandtl numbers of *k* (TKE) and ε (dissipation rate), tG_k is the generation of TKE due to the average velocity gradient, described by:

$$G_k = \mu_t S^2 \tag{6}$$

In which, μ_t is the turbulent dynamic viscosity, which are defined by:

$$\mu_t = \rho C_\mu \frac{k^2}{\varepsilon} \tag{7}$$

S is the modulus of the mean rate of strain tensor, described by:

$$S = \sqrt{2S_{ij}S_{ij}} \tag{8}$$

$$S_{ij} = \frac{1}{2}\left(\frac{\partial u_i}{\partial x_j} + \frac{\partial u_j}{\partial x_i}\right) \tag{9}$$

The coefficients appearing in the above equations are as follows: $C_1 = \max\left[0.43, \frac{\eta}{\eta+5}\right]$, $\eta = S\frac{k}{\varepsilon}$, σ_k = 1.0, σ_ε = 1.2, C_2 = 1.9.

2.2. Geometric Model and Boundary Conditions

Under the action of ocean currents, the leaked crude oil from the wellhead will migrate and diffuse downstream. A two-dimensional numerical simulation method is used to capture the migration behavior and diffusion process of the spilled oil. The selected two-dimensional geometric calculation domain is a section parallel to the current direction. At the bottom of the calculation area is a crude oil leakage wellhead located on the seabed and perpendicular to the calculation area, and it leaks crude oil into the sea at a certain speed. The density of seawater increases unevenly with the increase of depth, but in this paper, it is assumed that the density of seawater does not change with the increase in depth. The oil spill wellhead is located 5 m from the left and 15 m from the right endpoint. The water depth is H = 20 m, and the wellhead size is D = 0.5 m. The overall diagram of the geometric calculation area and boundary conditions is shown in Figure 1.

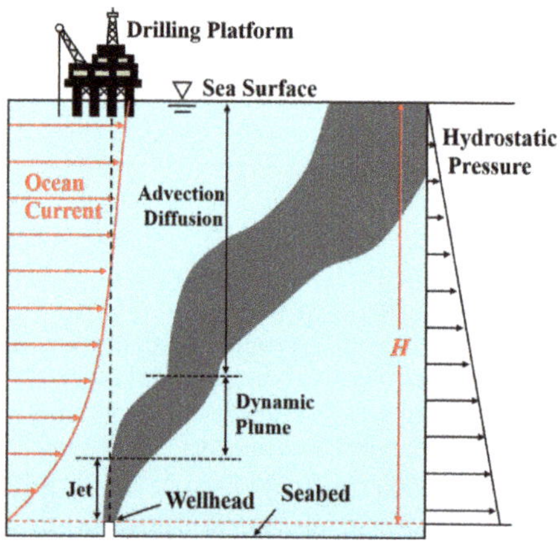

Figure 1. Underwater oil spill process.

The logarithmic speed profile is assumed for ocean currents to conform to the actual flow near the sea surface. The current velocity flow distribution is given as:

$$U = U_0\left[\frac{2y}{H} - \left(\frac{y}{H}\right)^2\right] \tag{10}$$

where U_0 is the maximum current velocity that appears at the free surface; in this paper, U_0 = 0.05~0.15 m/s is taken, and the maximum current velocity is located on the sea surface; H is the water depth, y is the height variable ($0 \leq y \leq H$).

The left side of the calculation domain and wellhead are set as the velocity-inlet boundary conditions. The velocity distribution of the ocean current meets the above velocity distribution relationship. The seabed and well wall at the bottom are set as the wall boundary conditions, and the other boundaries are selected as the outflow boundary conditions.

The governing equations are discretized by the finite volume method and solved by the Semi-Implicit Pressure Linked Equation (SIMPLE) algorithm. The time step was 0.001 s, the total number of time steps was set as 4×10^4, and the total calculation time was 40 s. The convergence criterion is 10^{-5} for the residual error of governing equations.

2.3. Grid Details and Grid Mesh Independence

The computational domain is meshed with structured hexahedral mesh elements, and the mesh generation near the wellhead is denser, which can more accurately describe the leakage state. The hexahedral mesh generation method based on topological segmentation is adopted to obtain the topological segmentation method of the model and carry out all hexahedral mesh generation. In the computational domain far from the wellhead, the mesh division is sparse, reducing the calculation time while ensuring the results were accurate.

A comparison of the rising altitude of spilled oil among grid densities and the difference in the rising height between 188,702 and 285,902 hexahedral nodes is less than 0.5% (as shown in Table 1). The grids with 188,702 hexahedral grids produce grid-independent results per the accuracy selected to balance computational cost and accuracy. The schematic diagram of the grid details is shown in Figure 2.

Table 1. Sensitivity of grid size to simulation results.

Mesh Numbers	Rising Height of Spilled Oil (m)	Error (%)
54,302	17.48	/
111,502	18.57	6.24%
188,702	18.83	1.40%
285,902	18.91	0.42%

Figure 2. Computational grid and solution domain.

2.4. Model Validation

The model's accuracy was verified according to the underwater oil spill experiment results of Zhu et al., as shown in Figure 3. The modeling method adopted in this paper was used to establish a matching numerical model regarding the experimental geometric conditions and oil physical properties. As seen in Figure 3, the simulation results based on the VOF model were in good agreement with the experimental results. Therefore, the numerical results were reliable and can be used to study the diffusion law of underwater oil spills.

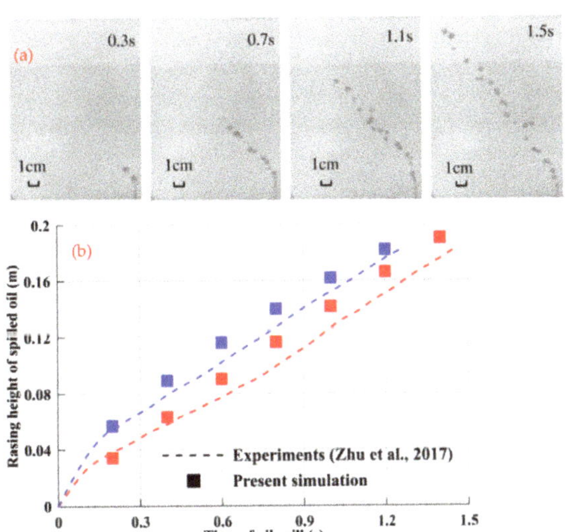

Figure 3. Validation of CFD model against experimental data [11]. (**a**) oil spill experiment; (**b**) CFD results against experimental data.

3. Results and Discussion

3.1. Effects of the Oil Spill Velocity

Firstly, the change of oil spill morphology with time at different oil spill velocities when $U_0 = 0$ m/s is obtained, as displayed in Figure 4. The time variable is shown horizontally, and the oil spill distribution at $T = 5$ s, $T = 10$ s, and $T = 15$ s is shown from left to right. Longitudinal is the variable of oil spill velocity; that is, each row represents the cloud map of oil spill concentration distribution when the oil spill velocity is 0.05 m/s, 0.10 m/s, 0.15 m/s, and 0.20 m/s, respectively, and the oil spill velocity increases from top to bottom.

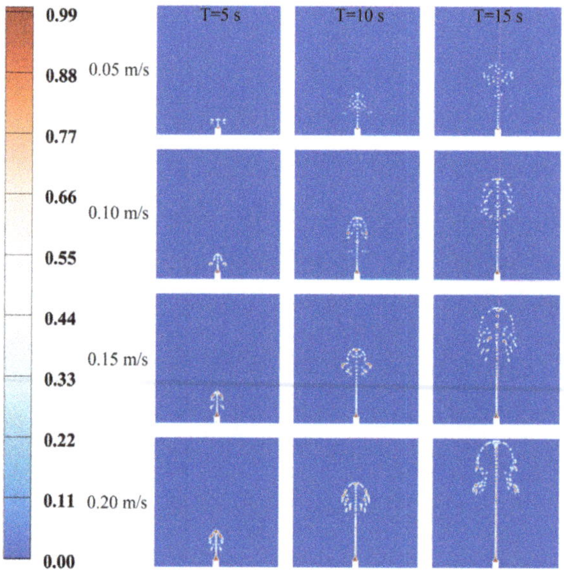

Figure 4. Analysis of influence of spill velocity on oil spill leakage and diffusion ($U_0 = 0$ m/s).

For the different initial oil spill velocities, the lateral migration distance of spilled oil does not change significantly with the increase of spill velocity in the absence of ocean current, which indicates that different spill velocities have no significant impact on the diffusion of the horizontal oil spill. Under different oil spill velocities, the changing trend of longitudinal distance is linear with spill velocity. When the crude oil leaves the wellhead, it is greatly affected by the initial momentum. Then, its acceleration gradually decreases as the oil droplets float upward. The migration of spilled oil mass can be divided into two successive stages: the accumulation stage and the buoyant droplet stage. In the floating process, the pressure difference is the main driving force in the initial stage, and the droplet buoyancy is the driving force in the other stages.

3.2. Effects of the Ocean Current Velocity

The ocean current velocity is the main factor affecting the diffusion of spilled oil. Ocean current can accelerate the dilution and purification of pollutants due to their rapid diffusion and correspondingly expand the scope of pollution. The change of oil spill diffusion morphology with time under different ocean current velocities is obtained when the oil spill velocity is $U_p = 0.15$ m/s, as shown in Figure 5. Among them, longitudinal is the time variable; that is, each line from top to bottom represents the oil–water distribution when the time point is $T = 10$ s, $T = 20$ s, $T = 30$ s, and $T = 40$ s, respectively; the horizontal is the maximum current velocity; that is, each column represents the oil spill distribution with the maximum current velocity of 0.05 m/s, 0.10 m/s, and 0.15 m/s, respectively.

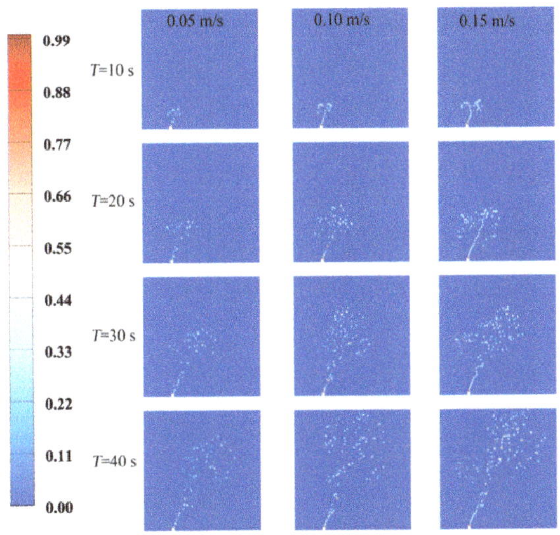

Figure 5. Effects of different current velocities on oil spill leakage and diffusion (U_P = 0.15 m/s).

It can be acknowledged that the length of the oil column increases with the increase of ocean current velocity. The momentum effect loses its dominant position when the oil column rises to a certain height. The positive buoyancy formed by the density difference controls the free diffusion of the oil droplets, and the oil column is blown into oil droplets under the action of the ocean current. Under the action of the ocean current, the angle between the oil column and horizontal plane changes with the change of ocean current velocity, mainly due to the oil column being easier to tilt along the current direction. According to the oil–water distribution corresponding to T = 30 s and T = 40 s, the momentum effect of the upper half of the oil column basically disappears, and only a small amount of oil droplets disperses. The reason is that when the oil column rises to a certain height, its initial momentum will be weakened, which will reduce its ability to resist ocean currents, thus forming a depression in the upper part.

The local velocity distribution of the underwater oil spill is shown in Figure 6. The three different current velocities correspond to different oil spill diffusion areas. When U_0 = 0.15 m/s, the corresponding diffusion range is the largest, and when U_0 = 0.05 m/s, the corresponding diffusion range is the smallest. The size of the diffusion region mainly depends on the number of dispersed oil droplets. Near the sea level, on the one hand, the initial momentum of the oil column decreases greatly; on the other hand, due to the large current velocity near the sea surface, the number of oil droplets is also large. When U_0 = 0.15 m/s, the crude oil will be affected by a strong current after being ejected from the wellhead. The upper half of the oil column will be blown away to form oil droplets, extending horizontally. At this time, the momentum of the oil column is not so significant, and it will be affected by the greater current velocity near the surface.

On this basis, the vorticity distribution characteristics of the spilled oil at T = 40 s are obtained, as shown in Figure 7. With the continuous overflow of crude oil from the wellhead, the oil spill trajectory has a centralized effect. The ocean current has a significant impact on the horizontal diffusion of the spilled oil, resulting in the dispersion effect of its motion trajectory. The main body of spilled oil inclines to the right and diffuses upward under the action of the transverse current, and a small part of the spilled oil diffuses to the left under the action of free drift; in the initial stage of the oil spill, the spilled oil is mainly distributed as dispersed oil droplets and oil blocks. With the increase of oil spill degree, a large number of continuous oil masses and oil belts also increase, the range of underwater pollution will also increase, and the more significant current velocity will also promote the growth of oil droplets.

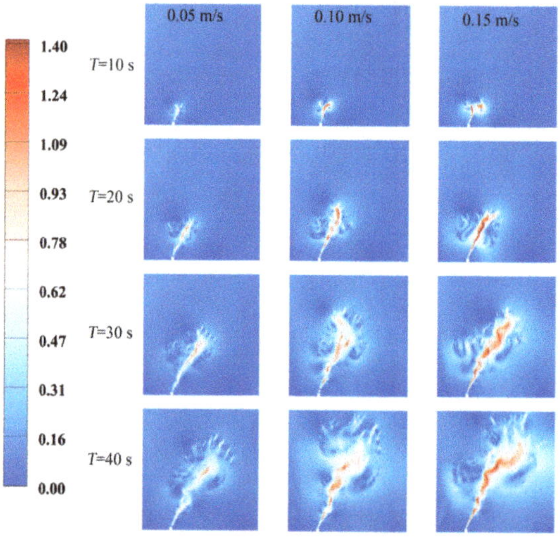

Figure 6. Influence of ocean current velocity on oil spill leakage and diffusion (U_P = 0.15 m/s).

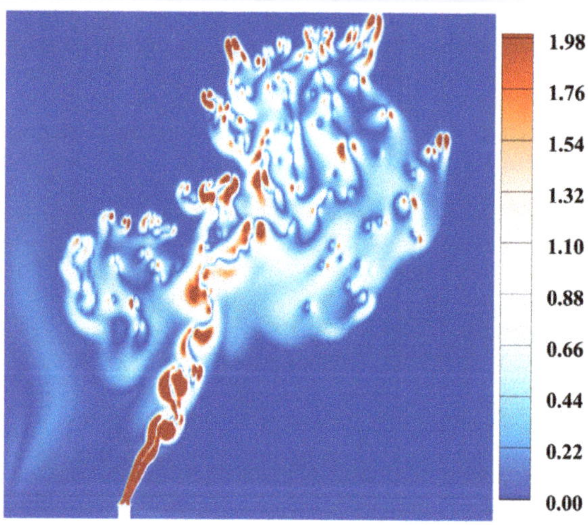

Figure 7. Contour plots of vorticity field influenced by ocean currents. (U_0 = 0.15 m/s, U_P = 0.15 m/s).

The rising height H_p under different maximum current velocities is compared, as shown in Figure 8. In the first 5 s, the spilled oil has approximately the same rising height because the oil spill leaked from the wellhead has relatively large inertia, which exceeds the shear effect of the ocean current. The crude oil is dispersed or combined under the joint action of gravity, inertia force, buoyancy, and shear stress, and its size changes with time. At the same time, with the growth of rising height, the viscous shear force gradually weakens the inertia force of the spilled oil. With the increase of H_p to 5 m, the shear effect of the ocean current begins to strengthen, and the greater the velocity of the ocean current, the more shear stress and kinetic energy it exerts on the spilled oil, thus the rising height of the spilled oil is slightly increased.

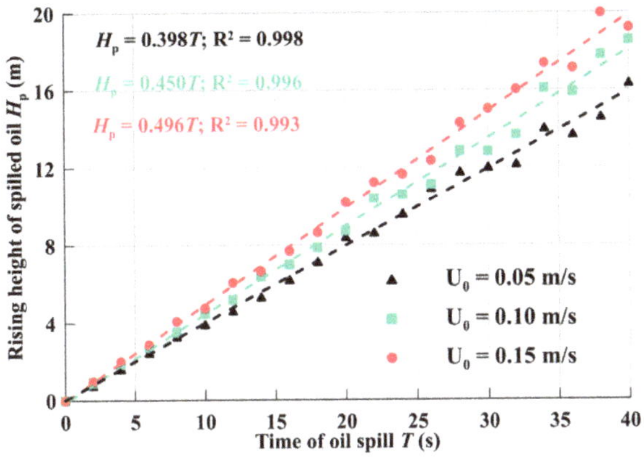

Figure 8. Rising height of spilled oil for three different maximum ocean current velocities. (U_P = 0.10 m/s).

The variation of lateral migration distance W_p with time under different current velocities is compared, as shown in Figure 9. The oil droplets move downstream by the shear action of the ocean current while floating. With the gradual increase of the maximum ocean current velocity, the lateral migration distance of the spilled oil gradually increases. When T = 10 s, the growth rate of the lateral migration distance of the spilled oil under the water is almost the same. In comparison, when T = 40 s, the lateral migration distance growth rate near the sea surface is significantly greater than its growth rate under the water. The crude oil moves downstream, and the lateral migration distance increases significantly with the decrease of water depth. The reason is that the upper seawater flow velocity is higher, exerting greater kinetic energy on the oil flow; therefore, the seawater flow plays an important role in the transportation and diffusion of crude oil. The impact of the actual seawater flow velocity distribution on the spilled oil trajectory should be considered, and a relatively accurate prediction should be made.

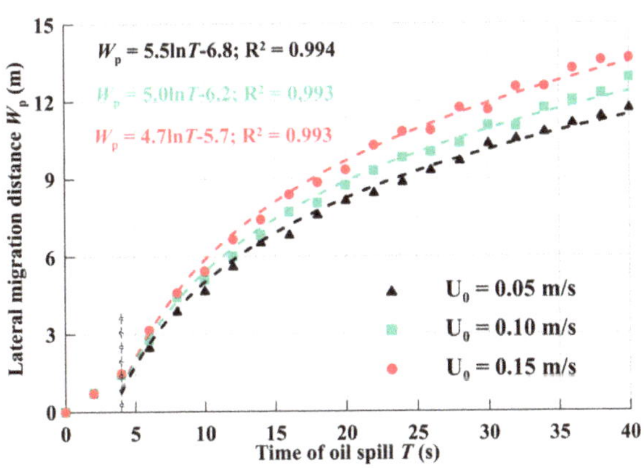

Figure 9. Lateral migration distance of spilled oil for three different maximum ocean current velocities. (U_P = 0.10 m/s).

3.3. Effects of the Crude Oil Density

The variation of the rising height H_p under different crude oil densities is obtained, as shown in Figure 10. The crude oil density affects the diffusion speed and diffusion range in seawater. The greater the density, the longer it takes crude oil to diffuse from the seabed to the sea surface. When ρ_P = 750 kg/m^3, the leaked oil diffuses from the leaking wellhead to the sea surface in about 28 s. When ρ_P = 900 kg/m^3, the rising distance of the spilled oil is 9.7 m at T = 28 s, which is 10.3 m different from the crude oil with a density of ρ_P = 750 kg/m^3 at the same time. This is because the higher the density of crude oil with the same volume, the greater the gravity and the same buoyancy. Therefore, the vertical upward force on high-density crude oil is smaller, and the rising speed is slower.

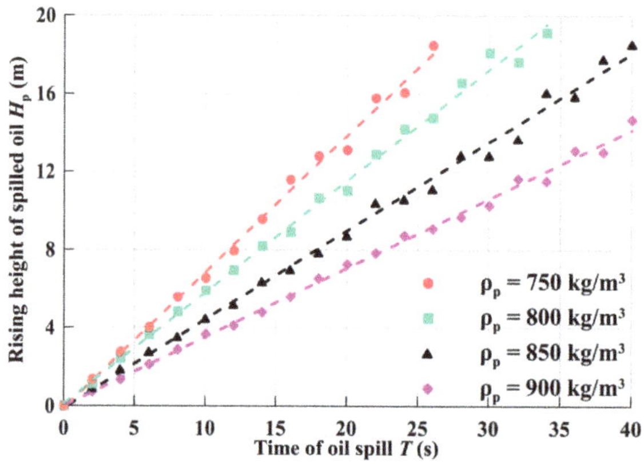

Figure 10. Rising height of spilled oil for three different oil densities. (U_P = 0.10 m/s).

The lateral migration distance W_P with time under different crude oil densities is obtained, as shown in Figure 11. An increase in the density of crude oil leads to a decrease in lateral migration distance in seawater. When ρ_P = 750 kg/m^3, the lateral migration distance of leaked oil in seawater after about 28 s is 14.7 m; when ρ_P = 900 kg/m^3, the lateral migration distance of spilled oil at T = 28 s is 8.5 m, which is 6.2 m different from the crude oil with a density of ρ_P = 750 kg/m^3 at the same time. Before the spilled oil reaches the sea surface, it is challenging to detect the lateral movement under the sea surface with monitoring instruments. In addition, floating containment booms are an essential device in the fight against coastal pollution, allowing us to contain the pollutant before its recovery [21]. Therefore, studying the maximum horizontal movement of crude oil with different densities is of great significance.

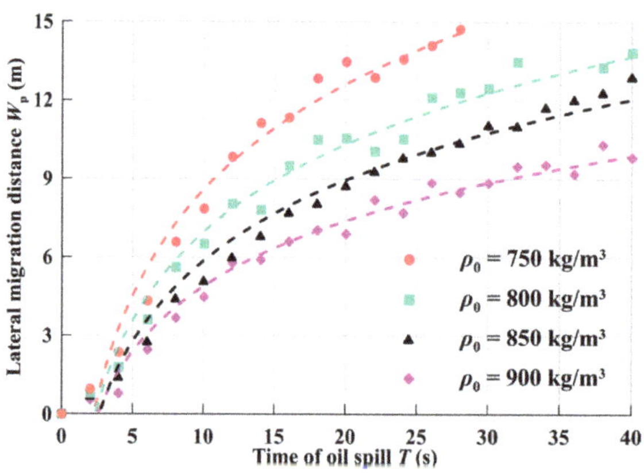

Figure 11. Lateral migration distance of spilled oil for three different oil densities. (U_P = 0.10 m/s).

3.4. Effects of the Crude Oil Viscosity

The viscosity is an essential quantitative parameter to characterize fluid rheology and can indicate the difficulty of crude oil flow. The viscosity change will change the diffusion law of spilled oil underwater. Therefore, the variation of the rising height H_p with time under different viscosities is obtained, as shown in Figure 12. The smaller the viscosity of the leaked crude oil, the faster the oil leakage rises from the leaking wellhead to the sea surface, and the shorter the time. When crude oil viscosity is $\mu = 0.01$ Pa.s at $T = 30$ s, the rising distance of the spilled oil is about 15.1 m, and when the crude oil viscosity is $\mu = 0.50$ Pa.s, the rising distance is about 12.5 m.

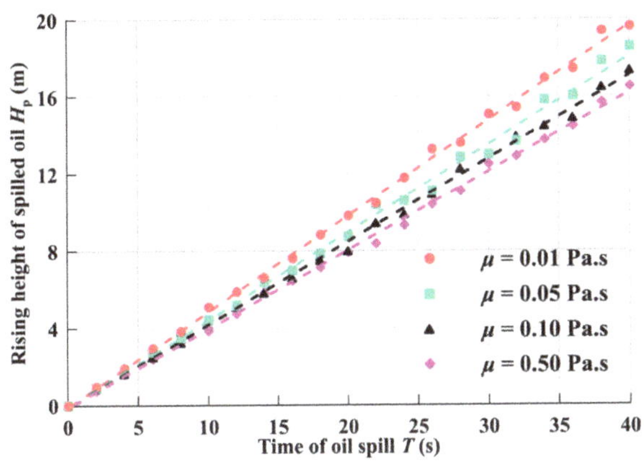

Figure 12. Rising height of spilled oil for three different oil viscosities.

The variation of crude oil lateral drift distance W_p with time under different crude oil viscosities is obtained, as displayed in Figure 13. The greater the viscosity of crude oil, the smaller the lateral diffusion distance in seawater. At $\mu = 0.01$ Pa.s, the lateral migration distance of the leaked oil in the seawater after about 28 s is 13.6 m, when $\mu = 0.50$ Pa.s at $T = 28$ s, the lateral diffusion distance of the spilled oil is 11.1 m, and the viscosity at the same time is $\mu = 0.01$ Pa.s and crude oil has a difference of 2.5 m. The increase of the viscosity will reduce the diffusion speed and drift speed of the spilled oil, and the oil mass with high viscosity will also block the drift and dispersion of the spilled oil. Therefore, the thickness of the oil film in the high-viscosity spilled oil is relatively thick; The slower diffusion speed provides more response time for oil spill emergency response.

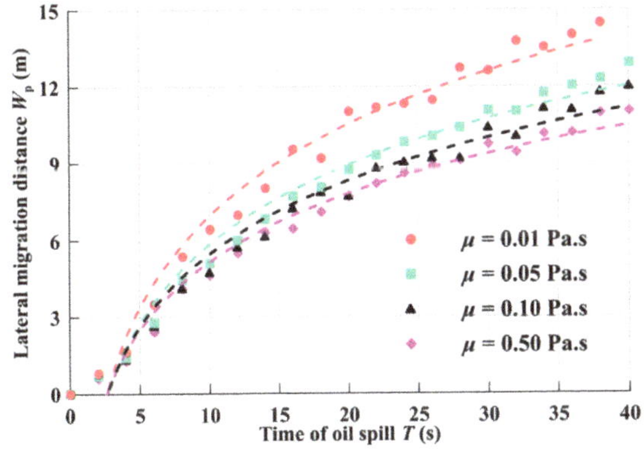

Figure 13. Lateral migration distance of spilled oil for four different oil viscosities.

3.5. Parameter Fitting

According to the above analysis, the lateral migration distance W_p is an important index to determine the impact range of an oil spill, so it is necessary to provide the variation of W_p with oil spill time under different working conditions. According to the numerical simulation results, the lateral migration distance W_p at different times under various working conditions is obtained, and the parameters are fitted, as seen in Equation (11). It is satisfied between the oil spill time and various influencing factors, and the power relationship between the oil spill lateral migration distance W_p and various influencing factors.

$$W_p = 50.8\mu^{0.42}\rho_0^{-0.45}\left(U_0 U_p\right)^{0.98}T^{1.25} \quad (11)$$

Figure 14 shows fitted curves of the relationship between the simulated lateral migration distance and the model-predicted lateral migration distance. Our correlations reasonably agreed, within ±15%, in comparison with the numerical lateral migration distance.

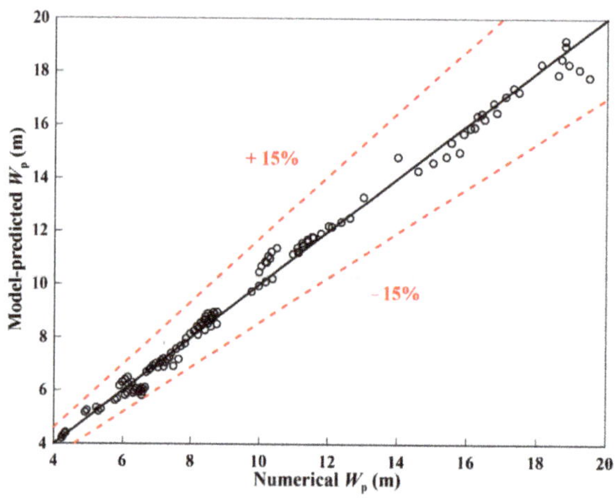

Figure 14. A comparison between the numerical results and the model-predicted results using the presented correlations.

3.6. Remediation Suggestions on Oil Spill Treatment Measures

Through the establishment of an emergency response mechanism for science, which can effectively protect the marine ecological environment, we propose to respond to underwater oil spill pollution. The results in this study show that it still takes some time for the oil droplets to reach the sea surface. So, improving the sensitivity, positioning accuracy, and response time of the oil spill early warning and detection device can maximize the emergency management time and reduce the damage of the oil spill accident to the environment. Emergency information construction can help people accurately judge the scope of oil spill accidents for the first time and judge the accident level and environmental damage risk.

In the buoyant droplet stage, the oil spill is greatly affected by the ocean current velocity, so it is necessary to strengthen the oil spill emergency information construction, such as real-time monitoring of the ocean current velocity change in the sea area where the drilling is located; real-time information processing, which can be used to guide the direction and scope of the sea area where the oil spill is spreading; and using unmanned aerial vehicle (UAV) and satellite monitoring technology to determine the sea area where the oil spill occurs in the shortest time, and take corresponding emergency treatment measures.

We propose strengthening the technical capacity of the emergency response team, including regular training and drills for emergency experts and professional emergency response teams, to improve the emergency management and disposal level of personnel;

improving the specialization of oil spill removal equipment, such as oil spill recovery ships, oil containment booms, oil collectors and oil spill collection and storage equipment; and minimizing the damage to ecology and the environment as much as possible.

4. Conclusions

Using the computational fluid dynamics (CFD) method and the volume-of-fluid (VOF) technique, the oil spill leakage and diffusion process was simulated. The effects of oil spill velocity, ocean current velocity, crude oil density, and crude oil viscosity were analyzed, and the rising height and lateral migration distance were obtained. The continuous oil flow is formed after the crude oil overflows from the wellhead. Due to the small amount of oil spilled per unit of time, after a while the continuous oil flow is interrupted by the ocean current and the seawater turbulence. Therefore, the spilled oil is soon distributed in the water as an oil mass. The current velocity is the main factor affecting the oil spill diffusion. With the increase of the oil spill height, the viscous shear force gradually weakens the oil spill's inertia force, and the current shear effect begins to increase. The greater the current velocity, the more shear stress and kinetic energy it exerts on the oil spill, so the oil spill's vertical and horizontal diffusion distance slightly increases. The higher the density of crude oil with the same volume, the greater the gravity. The change of oil spill viscosity will change the diffusion law of oil spill underwater. The greater the oil spill viscosity, the horizontal and vertical migration distance of crude oil will be slightly reduced, and the leakage and diffusion range of oil spill will be slightly reduced. The results of this study can provide a scientific basis for formulating an emergency treatment plan for offshore oil spill accidents and minimizing the harm to marine ecology and the environment. Later, we will discuss different ocean current densities and crude oil mixed with natural gas.

Author Contributions: Conceptualization, R.L.; Investigation, G.J.; Methodology, S.D.; Writing—original draft, S.D.; Writing—review and editing, R.L. and G.J. All authors have read and agreed to the published version of the manuscript.

Funding: This work was supported by the National Natural Science Foundation Project of China (Basic Research Program on Deep Petroleum Resource Accumulation and Key Engineering Technologies, Grant No. U19B6003).

Conflicts of Interest: The authors declare no conflict of interest.

References

1. Zhang, W.; Xie, P.; Li, Y.; Teng, L.; Zhu, J. Hydrodynamic characteristics and mass transfer performance of rotating packed bed for CO2 removal by chemical absorption: A review. *J. Nat. Gas Sci. Eng.* **2020**, *79*, 103373. [CrossRef]
2. Zidane, A.; Firoozabadi, A. Fracture-cross-flow equilibrium in compositional two-phase reservoir simulation. *SPE J.* **2017**, *22*, 950–970. [CrossRef]
3. Moreira, G.; Magalhães, H.L.F.; de Almeida Tavares, D.P.S.; de Brito Correia, B.R.; Leite, B.E.; da Costa Pereira, A.B.; de Farias Neto, S.R.; de Lima, A.G.B. Fluid Leakage in Submerged Offshore Pipeline: An Analysis of Oil Dispersion in Seawater. *Open J. Fluid Dyn.* **2020**, *10*, 95–121. [CrossRef]
4. Yang, Z.; Yu, J.; Li, Z.; Chen, H.; Jiang, M.; Chen, X. Application of computational fluid dynamics simulation for submarine oil spill. *Acta Oceanol. Sin.* **2018**, *37*, 104–115. [CrossRef]
5. Beyer, J.; Trannum, H.C.; Bakke, T.; Hodson, P.V.; Collier, T.K. Environmental effects of the Deepwater Horizon oil spill: A review. *Mar. Pollut. Bull.* **2016**, *110*, 28–51. [CrossRef]
6. Wei, L.; Hu, Z.; Dong, L.; Zhao, W. A damage assessment model of oil spill accident combining historical data and satellite remote sensing information: A case study in Penglai 19-3 oil spill accident of China. *Mar. Pollut. Bull.* **2015**, *91*, 258–271. [CrossRef]
7. Keramea, P.; Spanoudaki, K.; Zodiatis, G.; Gikas, G.; Sylaios, G. Oil spill modeling: A critical review on current trends, perspectives, and challenges. *J. Mar. Sci. Eng.* **2021**, *9*, 181. [CrossRef]
8. Omar, M.Y.; Shehada, M.F.; Mehanna, A.K.; Elbatran, A.H.; Elmesiry, M.M. A case study of the Suez Gulf: Modelling of the oil spill behavior in the marine environment. *Egypt. J. Aquat. Res.* **2021**, *47*, 345–356. [CrossRef]
9. Guo, W.; Zhang, S.; Wu, G. Quantitative oil spill risk from offshore fields in the Bohai Sea, China. *Sci. Total Environ.* **2019**, *688*, 494–504. [CrossRef]
10. Ye, X.; Chen, B.; Lee, K.; Storesund, R.; Li, P.; Kang, Q.; Zhang, B. An emergency response system by dynamic simulation and enhanced particle swarm optimization and application for a marine oil spill accident. *J. Clean. Prod.* **2021**, *297*, 126591. [CrossRef]

11. Zhu, H.; You, J.; Zhao, H. Underwater spreading and surface drifting of oil spilled from a submarine pipeline under the combined action of wave and current. *Appl. Ocean Res.* **2017**, *64*, 217–235. [CrossRef]
12. Kyaw, W.P.; Sugiyama, M.; Takagi, Y.; Suzuki, H.; Kato, N. Numerical analysis of tsunami-triggered oil spill from industrial parks in Osaka Bay. *J. Loss Prev. Process Ind.* **2017**, *50*, 325–336. [CrossRef]
13. Raznahan, M.; An, C.; Li, S.S.; Geng, X.; Boufadel, M. Multiphase CFD simulation of the nearshore spilled oil behaviors. *Environ. Pollut.* **2021**, *288*, 117730. [CrossRef]
14. Zhu, H.; Lin, P.; Pan, Q. A CFD (computational fluid dynamic) simulation for oil leakage from damaged submarine pipeline. *Energy* **2014**, *64*, 887–899. [CrossRef]
15. Zhu, H.; You, J.; Zhao, H. An experimental investigation of underwater spread of oil spill in a shear flow. *Mar. Pollut. Bull.* **2017**, *116*, 156–166. [CrossRef] [PubMed]
16. Sun, Y.; Cao, X.; Liang, F. Investigation on underwater spreading characteristics and migration law of oil leakage from damaged submarine pipelines. *Process Saf. Environ. Prot.* **2019**, *127*, 329–347. [CrossRef]
17. Ji, H.; Xu, M.; Huang, W.; Yang, K. The Influence of Oil leaking rate and Ocean Current Velocity on the Migration and Diffusion of Underwater Oil Spill. *Sci. Rep.* **2020**, *10*, 9226. [CrossRef] [PubMed]
18. Xu, W.; Wang, S.; Liu, G.; Zhang, Q.; Hassan, M.; Lu, H. Experimental and numerical investigation on heat transfer of Therminol heat transfer fluid in an internally four-head ribbed tube. *Int. J. Therm. Sci.* **2017**, *116*, 32–44. [CrossRef]
19. Xiong, J.; Cheng, R.; Lu, C.; Chai, X.; Liu, X.; Cheng, X. CFD simulation of swirling flow induced by twist vanes in a rod bundle. *Nucl. Eng. Des.* **2018**, *338*, 52–62. [CrossRef]
20. Yan, T.; Qu, J.; Sun, X.; Chen, Y.; Hu, Q.; Li, W.; Zhang, H. Numerical investigation on horizontal wellbore hole cleaning with a four-lobed drill pipe using CFD-DEM method. *Powder Technol.* **2020**, *375*, 249–261. [CrossRef]
21. Han, Y.; Nambi, I.M.; Prabhakar Clement, T. Environmental impacts of the Chennai oil spill accident—A case study. *Sci. Total Environ.* **2018**, *626*, 795–806. [CrossRef] [PubMed]

processes MDPI

Review

Microbial-Mediated Emissions of Greenhouse Gas from Farmland Soils: A Review

Han Wang [1], Rui Zhao [1], Dan Zhao [1], Shejiang Liu [1], Jianfeng Fu [1], Yuxin Zhang [2], Nan Dai [2], Dan Song [3] and Hui Ding [1,*]

[1] School of Environment Science and Engineering, Tianjin University, Tianjin 300072, China
[2] College of Materials Science and Engineering, Chongqing University, Chongqing 400044, China
[3] Chongqing Academy of Eco-Environmental Sciences, Chongqing 401147, China
* Correspondence: dinghui@tju.edu.cn

Abstract: The greenhouse effect is one of the concerning environmental problems. Farmland soil is an important source of greenhouse gases (GHG), which is characterized by the wide range of ways to produce GHG, multiple influencing factors and complex regulatory measures. Therefore, reducing GHG emissions from farmland soil is a hot topic for relevant researchers. This review systematically expounds on the main pathways of soil CO_2, CH_4 and N_2O; analyzes the effects of soil temperature, moisture, organic matter and pH on various GHG emissions from soil; and focuses on the microbial mechanisms of soil GHG emissions under soil remediation modes, such as biochar addition, organic fertilizer addition, straw return and microalgal biofertilizer application. Finally, the problems and environmental benefits of various soil remediation modes are discussed. This paper points out the important role of microalgae biofertilizer in the GHG emissions reduction in farmland soil, which provides theoretical support for realizing the goal of "carbon peaking and carbon neutrality" in agriculture.

Keywords: microorganism; farmland soil; greenhouse gas emission; microalgae biofertilizer

1. Introduction

Greenhouse gases (GHGs) usually refer to gases that can absorb the Earth's thermal radiation and enhance the greenhouse effect, mainly, carbon dioxide (CO_2) and methane; the greenhouse effect is one of the important environmental problems humans have so far faced in the 21st century. CO_2 is the single most important anthropogenic GHG in the atmosphere, contributing approximately 66% of the radiative forcing by long-lived greenhouse gases (WMO, 2019). Soil is the largest carbon reservoir in the terrestrial ecosystem [1]; the global carbon storage in 1~3 m of soil is about 1500~2344 Gt C (1 Gt = 10^{15} g), which is about three times that of the global vegetation and two times that of the atmosphere (IPCC, 2013b). However, the respiration of microorganisms, animals and roots, and the oxidation of carbonaceous matter also produce CO_2 [2]. Not only does soil produce CO_2, but the consumption of diesel, gasoline and electricity in farmland practices such as farming, irrigation and harvesting also cause CO_2 emissions [3]. The annual global emission of CH_4 was about 580 million in 2021. CH_4 is the second most important GHG after CO_2, with an average lifetime of about 8.75 years in the atmosphere and a contribution rate of about 15% of the greenhouse effect. The warming effect of CH_4 per unit mass in 20 years is about 84~87 times that of CO_2, and its warming effect in 100 years is about 28~36 times that of CO_2 [4]. The main emission sources of CH_4 in agriculture are rice and livestock cultivation, and the anaerobic environment of flooded rice fields and animal intestine create favorable conditions for CH_4 production by methanogens. The main sources of CH_4 are natural wetlands, human activities and biomass burning, and tropical regions with high CH_4 emissions contribute 80% of global CH_4 emissions [5]. N_2O is another noteworthy GHG, accounting for about 7.9% of the greenhouse effect. Its average lifetime in the atmosphere is 114 years, and its global warming potential (GWP) is 296~310 times that of CO_2, which is the main destroyer of stratospheric ozone [6]. Nitrogen fertilizer application in agriculture is the main source of N_2O, and N_2O emissions caused by fertilization account for about 30% of global land emissions. Therefore, reducing N_2O emissions from farmland soil is urgent to alleviate the greenhouse effect [7–9].

At present, researchers have developed some measures and technologies for GHG emission reduction in farmland soil, mainly including adding biochar, returning straw to the field, and applying organic fertilizer or microalgae biofertilizer and soil improvers (such as lime and nitrification inhibitor, etc.) [10–14]. Microbes play a crucial role in the application of these mitigation measures and technologies. However, there are few reviews on the role of microorganisms in GHG emissions from farmland. This review summarizes the sources of CO_2, CH_4 and N_2O in farmland soils, and discusses the environmental impact factors of microorganisms in farmland soil GHG emissions and the GHG emission reduction mechanisms of microorganisms under different soil remediation modes.

2. GHG Production in Agroecosystem

The agroecosystem is an important source of CO_2. Agricultural processes generate 15 billion tons of CO_2 emission, accounting for 30% of global total emissions [15]. CO_2 emission from the soil is usually called "soil respiration", which is a process of metabolism of animals, roots, fungi and bacteria in the soil. It involves three biological processes (plant root respiration, soil microbial respiration and soil animal respiration) and one non-biological process (chemical oxidation of carbon-containing substances) [2,16]. CO_2 in the atmosphere converts into organic matter through the photosynthesis of plants, and then the carbon in the organic matter enters the soil in the form of root exudates, dead roots or fallen leaves. Under the action of soil microorganisms, it is transformed into soil organic matter and stored in the soil, forming soil carbon sink.

The agroecosystem is an important source of CH_4 emissions, accounting for 15~30% of the total emissions [17]. In soil with poor aeration, low carbon organic acids, H_2, CO_2 and other substances formed by the fermentation of other microorganisms generate CH_4 under the action of methanogens. CH_4 in the agroecosystem can be generated in two ways: (1) organic acids in the soil environment or the degradation products of organic acids, CO_2 and H_2 generate CH_4 under the action of methanogens, or methanogens use formic acid and CO to form CH_4; (2) the demethylation of methyl compounds under the action of methanogens to produce CH_4. Methane-oxidizing bacteria account for the largest proportion in dryland soil with good aeration. About 82% of CH_4 is absorbed and utilized by methane-oxidizing bacteria in the soil before being discharged into the atmosphere, and then entering the soil ecosystem [18,19].

N_2O discharged from farmland soil is mainly a by-product of microbial nitrification and denitrification, in which nitrification is divided into autotrophic nitrification and heterotrophic nitrification, and autotrophic nitrification is divided into two stages: (1) ammonia oxidation stage: ammonia-oxidizing archaea (AOA) and bacteria (AOB) first oxidize NH_3 to NH_2OH and then reduce it to NO_2^-; (2) nitrite oxidation stage: NO_2^- is oxidized to NO_3^- by nitrite-oxidizing bacteria. Heterotrophic nitrification is the transformation of organic ammonia nitrogen into NO_2^- and NO_3^- by nitrifying bacteria and fungi in an aerobic environment [20]. Denitrification is a process in which microorganisms reduce NO_3^- and NO_2^- to NO, N_2O and N_2 in the presence of anaerobic environment and various enzymes [21]. When the atmospheric pressure and soil moisture content change, N_2O in the atmosphere will enter the soil pores through physical diffusion, and the water and solution in the soil will also dissolve N_2O in the atmosphere, thus, introducing N_2O into the agroecosystem [22]. The production process of CO_2, CH_4 and N_2O in farmland soil is shown in Figure 1.

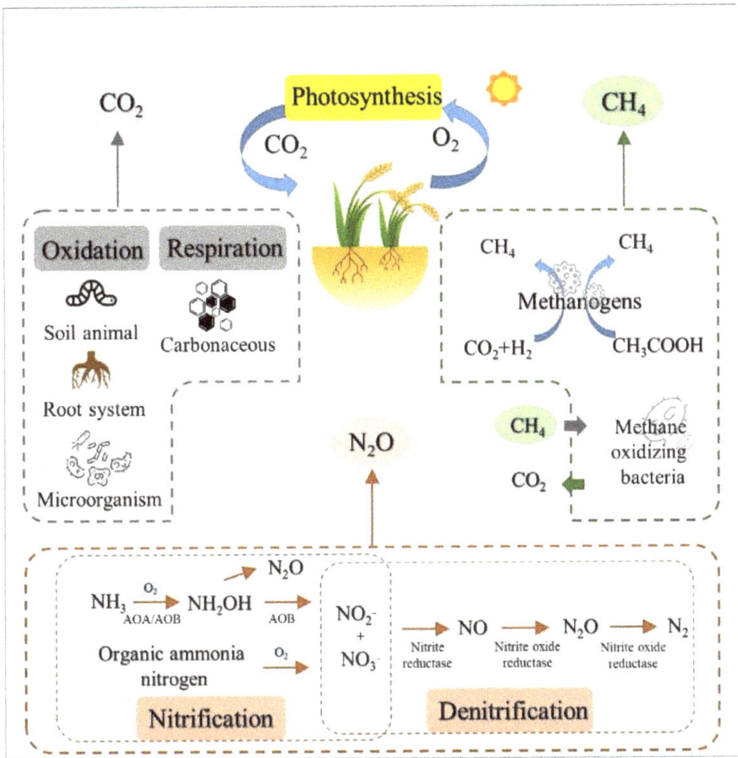

Figure 1. The production process of CO_2, CH_4 and N_2O in farmland soil.

3. Environmental Factors Affecting GHG Emissions of Microorganisms in Farmland Soil

3.1. Soil Temperature

A large number of studies have shown that temperature is the main factor affecting the production and emissions of GHG in soil [23–25]. Higher soil temperature can enhance the root respiration of crops, accelerate the decomposition of organic matter in the soil, improve the activity of microorganisms in the soil, and thus, accelerate the diffusion of CO_2 in soil [26,27]. Methanogens and methane-oxidizing bacteria jointly determine the emission of CH_4. Within a certain temperature, the metabolic capacity of methanogens is positively correlated with temperature. When the ambient temperature rises from 20 °C to 35 °C, the emission of CH_4 will double. However, recent studies have proposed that methanogens have thermal adaptability, and microbial activity decreases after long-term warming and increases after long-term cooling [28]. Walker et al. explored the response mechanism of soil microorganisms to temperature changes through in situ natural warming experiments, and the results showed that microbial temperature sensitivity and substrate consumption jointly affected soil carbon loss by controlling microbial biomass [29]. In the experiment of soil transplantation on a 3000 m elevation gradient in a tropical forest, every 1 °C increase in temperature resulted in a 4% decrease in soil carbon content. In addition to the decomposition of soil organic matter directly caused by temperature rise, temperature rise affected the physiological function of microorganisms, such as carbon utilization efficiency, microbial community change and the positive feedback effect of related enzyme activity [30]. The production of N_2O in soil has biological and abiotic pathways, and high temperature will stimulate microbial activity. Cui et al. conducted a liquid culture of *Pseudomonas mandelii* at 10~30 °C and found that its denitrification activity is proportional to the temperature [31]. Studies have found that high temperature has a more significant effect on the production of N_2O by abiotic pathways, and abiotic denitrification at 50 °C has the strongest effect and the highest N_2O emission [32].

3.2. Soil Moisture

Soil moisture affects the emissions of GHG from soils by changing the microbial activity and soil porosity in the soil itself. Soil CO_2 emission shows a Birch effect with soil moisture: a certain water content stimulates microbial activity and increases CO_2 emission, while too high a water content inhibits soil respiration [33]. Zou et al. studied GHG emissions patterns under different hydrological conditions and found that CO_2 equivalent emissions were the lowest when the groundwater level was close to the surface [30]. The utilization rate of O_2, the activity of microorganisms and the diffusion ability of gas molecules in soil are all affected by soil water content. Soil with a high water content is prone to form anaerobic areas, which promote the growth of methanogens and denitrification, thus, resulting in an increased emission of CH_4 and N_2O in soil [34,35]. The Paddy field is the main place where CH_4 and N_2O are produced, and its irrigation mode significantly affects the GHG emissions of the soil. Intermittent irrigation will inhibit the activity of methanogenic bacteria and reduce CH_4 emissions. Flooding irrigation provides an anaerobic environment to promote denitrification, but excessive water delays the diffusion of N_2O, resulting in the reduction of N_2O to N_2 and N_2O emission being reduced [36]. According to the above, the International Paddy Field Research Institute put forward the water management mode of dry–wet alternation; the lower redox potential of flooded soil is beneficial to the production of CH_4, and the higher redox potential of drained soil is beneficial to the production of N_2O [37]. By controlling the time of wet–dry alternation, the redox potential of soil is maintained at a moderately low level, and the lowest emissions of GHG is achieved [38]. Liao et al. found that soil moisture and atmospheric temperature will affect N_2O emission by adjusting the balance between nitrifying bacteria and denitrifying bacteria [36]. Soil moisture decreases with the increase in atmospheric temperature, which increases the gene abundance of *amoA* encoding nitrification to produce N_2O in soil, and the decrease in gene abundance of *nosZ* (Recombinant Nitrous-oxide reductase) encoding N_2O reduction to N_2, thus, increasing N_2O emission [39].

3.3. Soil Organic Matter

Soil organic matter (SOM) generally refers to a type of polymer organic compound with complex components and stable properties formed by organic residues in the soil through microorganisms or other physical and chemical processes [40]. SOM is a major carbon source for soil respiration and significantly affects soil GHG. Soil-activated organic C is a substrate for microbial growth, and its content directly affects the activity of microorganisms, which in turn affects the emission of GHG. Soluble organic matter content in SOM is closely related to CO_2 production in soil [41,42], and Paré and Bedard et al. found that alkane carbon and aromatic substances in the arctic tundra ecosystem enhanced CO_2 emission [43]. Wang et al. and Pascual et al. found that amines and aromatic compounds in the soil increased significantly after straw returned, resulting in higher CO_2 emissions [44,45]. SOM is the main substrate of methanogens, and the SOM content is positively correlated with the CH_4 emission [46]. Most denitrifying bacteria are chemoheterotrophic, and their energy for production and reproduction mainly comes from soil organic matter. Therefore, a high organic matter content provides sufficient energy for denitrification and promotes the production of N_2O [47]. Other studies have found that microorganisms decompose organic matter and consume oxygen, inhibit soil nitrification and reduce N_2O emission [48]. That this is due to the C/N in the soil directly affects the decomposition of SOM and the activity of microorganisms, thereby inhibiting or promoting the emission of N_2O from the soil [49].

3.4. Soil pH

The activities of microorganisms and enzymes in soil, the decomposition of organic matter and the development of crop roots are closely related to soil pH. The influence of soil pH on CO_2, CH_4 and N_2O emissions is complex. The optimum pH of most microorganisms in soil is 6–8, and too acidic or too alkaline an environment will inhibit the activity of microorganisms and reduce the emissions of GHG. The optimal pH for the growth and reproduction of methanogens is about 7 [50], and the acidic environment will reduce the emission of soil CH_4. N_2O reductase is the only enzyme that converts N_2O into N_2 during denitrification. Acidic soil will inhibit its activity or even cause its inactivation. Studies have found that the emission of N_2O in neutral soil is significantly lower than that in acidic soil [51,52]. Therefore, adding alkaline amendments such as $CaMg(CO_3)_2$, $CaCO_3$, $Ca(OH)_2$, CaO and other lime materials in acidic soil is beneficial to improve the activity of N_2O reductase and reduce the emission of N_2O [53]. Shaaban et al. and Wu et al. modified acidic soil with dolomite under different water gradients, and the results showed that soil pH increased rapidly after dolomite application, which promoted the conversion of N_2O into N_2 and reduced N_2O emission [54,55]. Shaaban et al. found that the concentration of NH_4^+-N decreased rapidly with time, while the concentration of NO_3^--N gradually increased after lime material was added to acidic soil, which indicated that nitrification in soil was strengthened [56]. The microorganisms consumed N_2O as an electron acceptor instead of NO_3^--N at higher NO_3^--N concentrations. A large amount of N_2O is converted into N_2 under the action of microorganisms, thereby reducing the emission of N_2O [56].

4. Microbial-Mediated Soil Emissions Reduction Mechanism under Different Soil Remediation Modes

4.1. Biochar

Biochar is a loose and porous substance with a high carbon content produced by carbonization organic materials under the condition of little or no oxygen. It has the characteristics of wide source, low cost, large specific surface area, strong adsorption capacity and strong carbon stability. Biochar can improve soil fertility and increase crop yield in agricultural applications. It has reportedly shown great potential in reducing GHG emissions in soils. A large number of experiments have found that fresh biochar cannot reduce CO_2 emission in soil [57–60], while biochar has been naturally aged in field soil, and the organic and inorganic complexes that accumulate on the surface of soil minerals can stabilize the organic carbon in biochar, structurally increasing spatial resistance and reducing CO_2 emissions from a physicochemical perspective [61]. In addition, compared with fresh biochar, aged biochar has a richer microbial community structure [62], and some CO_2-fixing bacteria appear, which reduces CO_2 emission on the microbial level [63].

The reduction in CH_4 emission by biochar is due to the joint action of physical chemistry and microorganisms in the soil. The application of biochar increases soil aeration and redox potential, which results in the reduction in CH_4 emission by physical–chemical reaction. Methanogens are obligate anaerobic bacteria, which are the main microorganisms producing CH_4 in the soil. After entering the soil, biochar with high porosity inhibits the activity of most methanogens and affects the change of microbial community in the soil [64]. Wang et al. monitored the microbial community after biochar application in soil for four consecutive years; the experimental results showed that the abundance of methanogens in the soil after long-term biochar application significantly decreased, while the abundance of methane-oxidizing bacteria did not change significantly, thus, reducing the emission of CH_4 in paddy fields [65].

The short-term addition of biochar to rice soil increased the abundance of ammonia-oxidizing bacteria (AOB) and ammonia monooxygenase gene (*amoA*), and significantly increased the denitrification rate of the soil. Fresh biochar provided a stronger alkaline environment and nutrients, and even improved the denitrification capacity and nitrogen emission [66]. Many studies have shown that fungi make a greater contribution to N_2O production than bacteria in acid soil [67,68]. As the denitrification product of fungi is

N_2O instead of N_2, reducing the number of fungi in soil can reduce N_2O emissions. Adding biochar and nitrogen fertilizer to acid soil with high N_2O emission will increase the soil pH, change the community composition of fungi, inhibit the denitrification of fungi, significantly reduce the abundance of fungi, increase the abundance of the *nosZ* gene, enhance the activity of N_2O reductase, and promote bacteria to reduce N_2O to N_2 [69] (Figure 2). *nosZ I* and *nosZ II* are N_2O reductase coding genes widely existing in the environment. Studies have shown that microbes containing the *nosZ II* gene have greater N_2O reduction potential. Some microbes containing the *nosZ II* gene lack the nitrite reductase gene, so they do not produce N_2O during denitrification, which provides a new research idea for N_2O emission reduction in the future.

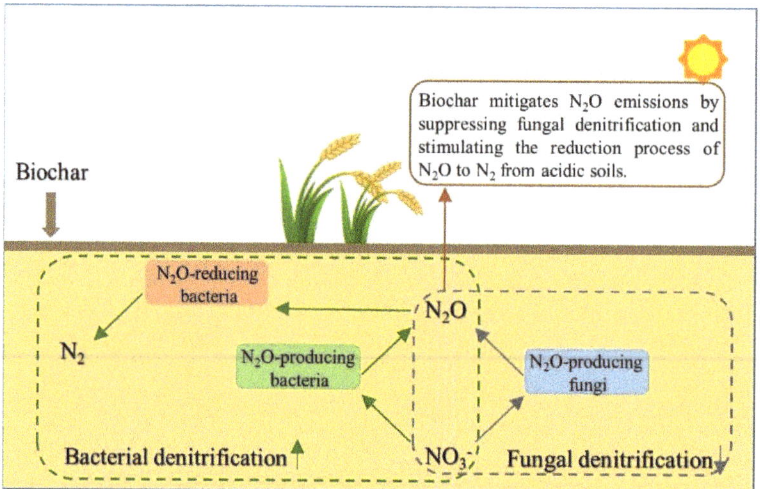

Figure 2. Potential mechanism of N_2O emission reduction following biochar amendment from acidic soils.

Although biochar can improve carbon sequestration, achieve emission reduction and adjust the abundance and activity of microorganisms related to GHG emissions in soil, it also has the health risk of releasing heavy metals, organic pollutants, nanoparticles and other substances to inhibit the growth and development of crops. Nanoparticles extracted from six biochars by Zhang et al. were confirmed to inhibit the germination of rice seeds and the growth of reed roots [70]. After biochar enters the soil, soil alkalinity will be enhanced, which will reduce the utilization rate of trace elements such as Fe, Zn and Cu in the soil, interfere with crop growth and even cause plant death [71]. Some studies have found that pollutants in biochar cause serious harm to earthworms [72], and excessive biochar directly reduces their survival rate [73]. Therefore, the application of biochar needs to be considered in combination with the actual soil environment, nature and other factors.

4.2. Organic Fertilizer

Organic fertilizer is the best substitute for chemical fertilizer by using agricultural, animal husbandry and industrial wastes as raw materials to turn waste into treasure. Organic fertilizer can significantly improve soil quality, enrich the microbial community and increase crop yield. However, studies have shown that the introduction of organic fertilizer into the soil will increase the content of light component organic carbon, which is more easily used by microorganisms, and the application of organic fertilizer alone will significantly increase soil CO_2 emission [74]. Wang et al. and Li et al. adopted the mode of fertilizer reduction combined with organic fertilizer application and found that soil carbon sequestration significantly increased and GHG emissions significantly decreased in double-cropping rice fields [75,76]. Studies have shown that CH_4 effluxes were significantly and negatively related to *mcrA* and *pmoA* gene copy numbers, and positively related to *mcrA*/*pmoA*. Organic fertilizers provide substrates for methanogens and promote the production and emission of CH_4 [77,78]. Li et al. replaced a part of inorganic fertilizer with organic fertilizer in the soil, and five substitution rates including 0, 20%, 50%, 80%, and 100% and a no fertilizer control were evaluated on Chinese cabbage. Cylindrical PVC chambers were placed at the center of each plot on each sampling day at 9 a.m. to collect gas. They found that organic fertilizer could reduce the emission of N_2O, and the quality of the vegetables improved under the substitution rate of 20~50% [79]. In summary, the rational use of organic fertilizer can not only regulate C/N in the soil, thereby changing the dominant species of microorganisms in the soil, but also increase crop yield and alleviate the GHG effect. Therefore, significant experimentation and research are needed to find the best case.

4.3. Straw Returning

Straw returning is a comprehensive utilization measure widely adopted around the world, which has the advantages of fertilizing soil capacity, improving cultivated land quality, and increasing soil carbon reservoir and crop yield. As an agricultural renewable resource, straw contains N, P, K, Ca, Mg and other mineral elements needed for crop growth. The main components of straw are abundant organic carbon such as cellulose, hemicellulose and lignin, which can improve the soil organic matter content after returning to the field. As shown in Table 1, there are differences in the composition of straw from different crops, which have different effects on GHG emissions in the soil after returning to the field. Zuo et al. studied the effect of returning corn straw pretreated with white rot fungi on soil GHG emissions, and the results showed that the emissions of CO_2 and N_2O increased significantly due to the increase in C and N content [11]. Recent studies have also suggested that straw return significantly increased the net GWP compared to non-straw return [80], which is consistent with the results of Wu et al., who reported that straw return increased GHG. Research on straw returning significantly increasing CH_4 emissions has been widely reported [81]. Wang et al. found that straw returning significantly increased CH_4 emissions by using the method of meta-analysis, and the comprehensive temperature potential of GHG significantly increased by 87.1% [82]. The impact of straw returning on N_2O is still uncertain. Li et al. and Liu et al. believed that straw returning increased the content of C in the soil, enhanced the denitrification of microorganisms in the soil, and promoted the emission of N_2O [83,84]. Xu et al. studied the impact of nitrogen fertilizer and straw on N_2O emission from winter wheat farmland. Four treatments, i.e., no N fertilizer and no straw, straw incorporation only, N fertilizer only, and N fertilization plus straw incorporation, were established in the experiment. They found that straw incorporation increased the N content in the soil but had no significant impact on N_2O emission [85]. Chen et al. used ^{15}N tracing technology to study the mechanism of N_2O increase after straw return [86]. They found that the C/N ratio of straw application was negatively related to soil denitrification, and increasing the C/N ratio of straw application could weaken the N_2O emission during denitrification. Straw returning significantly affects the soil microbial community structure, and the dominant bacteria in the straw degradation process will also change over time. In order to reduce GHG emissions, the strategy of straw incorporation should be adjusted. There is a research gap in the impact of straw return on GHG, which still needs to be studied by relevant professionals.

Table 1. The emission data of the GHGs from straw addition.

Soil Type	Straw Type	GHG	Compared with No Straw Addition	Year	Ref.
Rice-wheat	Rice-wheat	CH_4	+35.0%	2015	[87]
Rice-rapeseed	Rapeseed	CO_2 CH_4	+6.3% +32.9%	2016	[88]
Maize-crop	Maize straw	N_2O	−11.0~27.0%	2017	[89]
Rice paddy	Rice straw	CH_4 N_2O	+39.1% −77.8%	2017	[48]
Rice paddy	Rice straw	CO_2 CH_4 N_2O	+14.8~27.5% +36.9~182.1% −23.5~40.6%	2019	[90]
Rice-wheat	Wheat	CH_4	+36.6~80.1%	2021	[91]
Rice-wheat	Rice-wheat	CH_4 N_2O	+41.20% +47.50%	2021	[80]
Rice-wheat	Rice-wheat	CH_4 N_2O	+5.4~72.2% −3.3~31.4%	2021	[92]
Wheat	Wheat	CO_2 N_2O	+11.5~28.3% +37.1~48.4%	2022	[93]

4.4. Microalgae Biofertilizer

Microalgae are widely distributed unicellular or simple multicellular microorganisms in land, lake and sea. Microalgae can efficiently carry out photosynthesis and be used for energy production, wastewater treatment and CO_2 reduction. Microalgae biofertilizer is mainly composed of eukaryotic green algae with high photosynthetic efficiency and prokaryotic cyanobacteria with fixed nitrogen. Microalgae biofertilizer is rich in trace elements and has the advantages of high efficiency, environmental protection, carbon fixation and nitrogen fixation to reduce GHG emissions [94]. The beneficial effects of microalgae on soil and GHG are shown in Figure 3. The photosynthetic efficiency of microalgae is 10~50 times that of ordinary terrestrial plants. Microalgae can fix CO_2 from the atmosphere and increase O_2 content in the soil by absorbing CO_2 in the environment and releasing O_2 at the same time [95]. Microalgae in the soil can activate solidified phosphorus and potassium in soil under the action of biological enzymes, improve the activity of cationic mineral elements in soil, and promote the accumulation and transformation of photosynthetic products. The extracellular polysaccharides secreted by microorganisms and microalgae on the soil surface will form a layer of algal biofilm, which can increase the carbon and nitrogen sources in the soil by sequestering CO_2 and N_2 in the atmosphere [96]. Marks et al. added the suspension of chlorella culture to farmland soil, accelerating the formation of soil photosynthetic biofilm [97].

Cyanobacteria have both carbon and nitrogen-fixation functions. CO_2 in the atmosphere is fixed through photosynthesis, similar to green algae. The cyanobacteria are divided into vegetative and highly differentiated heterocyst cells. Heteroplasts have a unique nitrogenase, which can reduce N_2 to NH_3. Nitrate reductase and nitrite reductase in vegetative cells convert nitrate and nitrite in the environment to NH_3 through nitrification and denitrification, increasing soil nitrogen reserves [98]. Nitrogen-containing substances such as amino acids, sugars, polysaccharides and a small number of hormones secreted by cyanobacteria during their growth and reproduction further increase the content of effective nitrogen in the soil [99,100]. Ali et al. showed that the CH_4 emission flux of Bangladeshi rice soil treated with azolla and cyanobacteria was low in two consecutive rice experiments, 12% lower than that of the control [101]. Prasanna et al. conducted experiments in paddy fields in New Delhi, India, and found that the CH_4 emission of rice soil inoculated with two kinds of Anabaena biofilm (Anabaena—Trichoderma, and Anabaena—*Pseudomonas aeruginosa*) was 50~80% lower than that of rice fields under the traditional mode [102]. Shrestha et al. found that, compared with urea, microalgae biofertilizer did not significantly increase wheat yield, but reduced nitrogen oxide (N_2O and NO) emissions in soil [103].

Zhang et al. and Hu et al. tried to combine microalgae biofertilizer with biochar or organic fertilizer and found that the carbon sequestration ability of microalgae was significantly improved [104,105]. The reason for this is that the addition of biochar and organic fertilizer increases the intracellular glucose content of microalgae, and microorganisms are more likely to obtain extracellular glucose; thus, a large amount of intracellular glucose becomes a part of soil carbon sink, strengthening the carbon sequestration ability of microalgae. It has been reported that microalgal biofertilizer can not only sequester carbon, fix nitrogen and reduce GHG emissions, but the dead algal cells can be converted into organic matter and improve soil fertility and plant yield [106]. Microalgae carbon fixation is also widely used in the treatment of coal-fired flue gas in factories. Microalgae fix CO_2 in coal-fired flue gas through photosynthesis, and absorb NO_x and SO_x in flue gas as nitrogen and sulfur sources for their own growth and reproduction [107,108]. Microalgae, the product of industrial carbon fixation, happens to be an important source of microalgae biofertilizer, which will become an effective medium for industrial and agricultural carbon emissions reduction. Under the background of global green production, microalgae have broad application prospects and are important resources for future development.

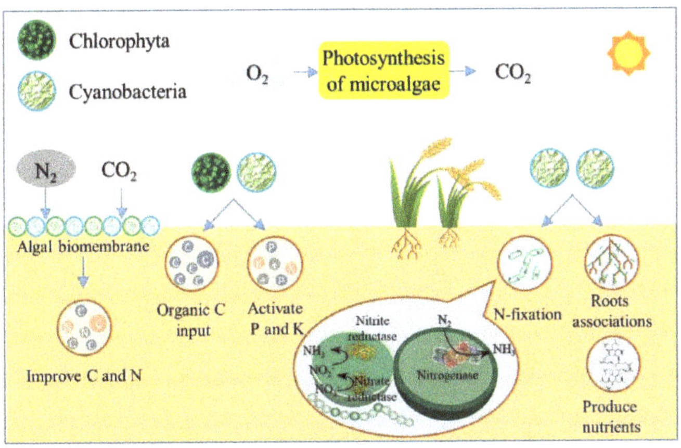

Figure 3. Beneficial effects of microalgae on soil and GHG emissions.

5. Conclusions and Prospects

There are many factors affecting GHG emissions from farmland soils. Soil temperature, soil moisture, soil organic matter content and soil pH, as well as other soil physical and chemical properties change soil GHG emissions by affecting the activities of soil microorganisms and related enzymes. At present, the feasible technologies to control soil GHG emissions include biochar application, organic fertilizer application, straw return and microalgae biofertilizer application. However, there are heavy metals, polycyclic aromatic hydrocarbons and other organic pollutants in biochar, which may inhibit crop growth, reduce crop yield and affect the growth and reproduction of soil animals after application; the organic fertilizer application and straw return require high operation and technology, so the emission reduction effect in actual application is not stable.

Here, we emphasize a remediation mode of "microalgae biofertilizer" with future development prospects. Microalgae biofertilizer satisfies people's demands for healthy soil; it achieves environmental protection, and agricultural quality and efficiency improvement through multiple functions such as carbon fixation and nitrogen fixation, crop growth promotion and soil improvement. However, there are few reports on the response mechanism of microorganisms in soil after applying microalgae biofertilizer. Therefore, it is significant to explore the underlying mechanism through the GHG emissions of soil after applying microalgae biofertilizer and the metagenome sequencing technology, which will provide important theoretical support for the development of microalgae biofertilizer.

Author Contributions: Conceptualization, H.W., R.Z. and H.D.; methodology, H.W. and D.Z.; software, H.W. and R.Z.; validation, H.W. and R.Z.; formal analysis, H.W. and D.Z.; investigation, H.W.; resources, H.W.; data curation, H.W. and R.Z.; writing—original draft preparation, H.W. and R.Z.; writing—review and editing, H.W. and R.Z.; visualization, H.W. and D.Z.; supervision, S.L., J.F., Y.Z., N.D., D.S. and H.D.; project administration, H.D.; funding acquisition, H.D. All authors have read and agreed to the published version of the manuscript.

Funding: This research was funded by the Scientific Research Project of Chongqing Ecological Environment Bureau (No. CQEE2022-STHBZZ118) and the Key Research and Development Plan of Tianjin (NO. 21YFSNSN00170).

Conflicts of Interest: The authors declare no conflict of interest.

References

1. Hao, Z.; Zhao, Y.; Wang, X.; Wu, J.; Jiang, S.; Xiao, J.; Wang, K.; Zhou, X.; Liu, H.; Li, J.; et al. Thresholds in aridity and soil carbon-to-nitrogen ratio govern the accumulation of soil microbial residues. *Commun. Earth Environ.* **2021**, *2*, 236. [CrossRef]
2. Oertel, C.; Matschullat, J.; Zurba, K.; Zimmermann, F.; Erasmi, S. Greenhouse gas emissions from soils A review. *Geochemistry* **2016**, *76*, 327–352. [CrossRef]
3. Deng, C.X.; Li, R.R.; Xie, B.G.; Wan, Y.L.; Li, Z.W.; Liu, C.C. Impacts of the integrated pattern of water and land resources use on agricultural greenhouse gas emissions in China during 2006–2017: A water-land-energy-emissions nexus analysis. *J. Clean. Prod.* **2021**, *308*, 127221. [CrossRef]
4. Ji, D.H.; Zhou, M.Q.; Wang, P.C.; Yang, Y.; Wang, T.; Sun, X.Y.; Hermans, C.; Yao, B.; Wang, G.C. Deriving Temporal and Vertical Distributions of Methane in Xianghe Using Ground-based Fourier Transform Infrared and Gas-analyzer Measurements. *Adv. Atmos. Sci.* **2020**, *37*, 597–607. [CrossRef]
5. Tian, H.; Chen, G.; Lu, C.; Xu, X.; Ren, W.; Zhang, C.; Zhang, B.; Banger, K.; Tao, B.; Pan, S.; et al. Global methane and nitrous oxide emissions from terrestrial ecosystems due to multiple environmental changes. *Ecosyst. Health Sustain.* **2015**, *1*, 1–20. [CrossRef]
6. Thompson, R.L.; Lassaletta, L.; Patra, P.K.; Wilson, C.; Wells, K.C.; Gressent, A.; Koffi, E.N.; Chipperfield, M.P.; Winiwarter, W.; Davidson, E.A.; et al. Acceleration of global N_2O emissions seen from two decades of atmospheric inversion. *Nat. Clim. Chang.* **2019**, *9*, 993–998. [CrossRef]
7. Cui, X.Q.; Zhou, F.; Ciais, P.; Davidson, E.A.; Tubiello, F.N.; Niu, X.Y.; Ju, X.T.; Canadell, J.G.; Bouwman, A.F.; Jackson, R.B.; et al. Global mapping of crop-specific emission factors highlights hotspots of nitrous oxide mitigation. *Nat. Food* **2021**, *2*, 886–893. [CrossRef]
8. Tian, H.Q.; Xu, R.T.; Canadell, J.G.; Thompson, R.L.; Winiwarter, W.; Suntharalingam, P.; Davidson, E.A.; Ciais, P.; Jackson, R.B.; Janssens-Maenhout, G.; et al. A comprehensive quantification of global nitrous oxide sources and sinks. *Nature* **2020**, *586*, 248–256. [CrossRef]
9. Syakila, A.; Kroeze, C. The global nitrous oxide budget revisited. *Greenh. Gas. Meas. Manag.* **2011**, *1*, 17–26. [CrossRef]
10. Zhang, B.; Zhou, M.; Lin, H.; Ntacyabukura, T.; Wang, Y.; Zhu, B. Effects of different long-term crop straw management practices on ammonia volatilization from subtropical calcareous agricultural soil. *Atmos. Ocean. Sci. Lett.* **2020**, *13*, 232–239. [CrossRef]
11. Zuo, S.S.; Wu, D.; Du, Z.L.; Xu, C.C.; Wu, W.L. Effects of white-rot fungal pretreatment of corn straw return on greenhouse gas emissions from the North China Plain soil. *Sci. Total Environ.* **2022**, *807*, 150837. [CrossRef]
12. Sun, J.; Haiyun, P.; Jianmin, C.; Xinming, W.; Min, W.; Weijun, L.; Lingxiao, Y.; Qingzhu, Z.; Wenxing, W.; Abdelwahid, M. An estimation of CO_2 emission via agricultural crop residue open field burning in China from 1996 to 2013. *J. Clean. Prod.* **2016**, *112*, 2625–2631. [CrossRef]
13. Chen, H.H.; Li, X.C.; Hu, F.; Shi, W. Soil nitrous oxide emissions following crop residue addition: A meta-analysis. *Glob. Chang. Biol.* **2013**, *19*, 2956–2964. [CrossRef]
14. Sun, D.Y.; Tang, X.F.; Li, J.; Liu, M.; Hou, L.J.; Yin, G.Y.; Chen, C.; Zhao, Q.; Klumper, U.; Han, P. Chlorate as a comammox Nitrospira specific inhibitor reveals nitrification and N_2O production activity in coastal wetland. *Soil Biol. Biochem.* **2022**, *173*, 108782. [CrossRef]
15. Terrer, C.; Phillips, R.P.; Hungate, B.A.; Rosende, J.; Pett, R.J.; Craig, M.E.; van Groenigen, K.J.; Keenan, T.F.; Sulman, B.N.; Stocker, B.D.; et al. A trade-off between plant and soil carbon storage under elevated CO_2. *Nature* **2021**, *591*, 599–603. [CrossRef]
16. Zheng, X.; Moses, A.A.; Hui, W.; Peng, H.; Shan, L.; Na, C.; Songyuan, W.; Hongling, Z.; Hui, D.; Kebin, L. Effect of lignin and plant growth-promoting bacteria (*Staphylococcus pasteuri*) on microbe-plant Co-remediation: A PAHs-DDTs Co-contaminated agricultural greenhouse study. *Chemosphere* **2020**, *256*, 127079. [CrossRef]
17. Saunois, M.; Stavert, A.R.; Poulter, B.; Bousquet, P.; Canadell, J.G.; Jackson, R.B.; Raymond, P.A.; Dlugokencky, E.J.; Houweling, S.; Patra, P.K.; et al. The Global Methane Budget 2000–2017. *Earth Syst. Sci. Data* **2020**, *12*, 1561–1623. [CrossRef]
18. Malyan, S.K.; Bhatia, A.; Kumar, A.; Gupta, D.K.; Singh, R.; Kumar, S.S.; Tomer, R.; Kumar, O.; Jain, N. Methane production, oxidation and mitigation: A mechanistic understanding and comprehensive evaluation of influencing factors. *Sci. Total Environ.* **2016**, *572*, 874–896. [CrossRef]
19. Mitra, S.; Majumdar, D.; Wassmann, R. Methane production and emission in surface and subsurface rice soils and their blends. *Agric. Ecosyst. Environ.* **2012**, *158*, 94–102. [CrossRef]

20. Zhong, L.; Bowatte, S.; Newton, P.C.D.; Hoogendoorn, C.J.; Luo, D. An increased ratio of fungi to bacteria indicates greater potential for N_2O production in a grazed grassland exposed to elevated CO_2. *Agric. Ecosyst. Environ.* **2018**, *254*, 111–116. [CrossRef]
21. Wang, C.; Amon, B.; Schulz, K.; Mehdi, B. Factors That Influence Nitrous Oxide Emissions from Agricultural Soils as Well as Their Representation in Simulation Models: A Review. *Agronomy* **2021**, *11*, 770. [CrossRef]
22. Muhammad, Z.M.; Edouard, M.; Alain, R.; Martial, B.; Alain, B. Xylophagous termites: A potential sink for atmospheric nitrous oxide. *Eur. J. Soil Biol.* **2012**, *53*, 121–125.
23. Lu, S.B.; Xu, Y.; Fu, X.P.; Xiao, H.; Ding, W.; Zhang, Y.J. Patterns and Drivers of Soil Respiration and Vegetation at Different Altitudes in Southern China. *Appl. Ecol. Environ. Res.* **2019**, *17*, 3097–3106. [CrossRef]
24. Kellman, L.; Myette, A.; Beltrami, H. Depth-Dependent Mineral Soil CO_2 Production Processes: Sensitivity to Harvesting-Induced Changes in Soil Climate. *PLoS ONE* **2015**, *10*, e0134171. [CrossRef]
25. Wang, Q.K.; Zhao, X.C.; Chen, L.C.; Yang, Q.P.; Chen, S.; Zhang, W.D. Global synthesis of temperature sensitivity of soil organic carbon decomposition: Latitudinal patterns and mechanisms. *Funct. Ecol.* **2019**, *33*, 514–523. [CrossRef]
26. Carey, J.C.; Tang, J.W.; Templer, P.H.; Kroeger, K.D.; Crowther, T.W.; Burton, A.J.; Dukes, J.S.; Emmett, B.; Frey, S.D.; Heskel, M.A.; et al. Temperature response of soil respiration largely unaltered with experimental warming. *Proc. Natl. Acad. Sci. USA* **2016**, *113*, 13797–13802. [CrossRef]
27. Voigt, C.; Lamprecht, R.E.; Marushchak, M.E.; Lind, S.E.; Novakovskiy, A.; Aurela, M.; Martikainen, P.J.; Biasi, C. Warming of subarctic tundra increases emissions of all three important greenhouse gases-carbon dioxide, methane, and nitrous oxide. *Glob. Chang. Biol.* **2017**, *23*, 3121–3138. [CrossRef]
28. Chen, H.; Zhu, T.; Li, B.; Fang, C.; Nie, M. The thermal response of soil microbial methanogenesis decreases in magnitude with changing temperature. *Nat. Commun.* **2020**, *11*, 5733. [CrossRef]
29. Walker, T.W.N.; Kaiser, C.; Strasser, F.; Herbold, C.W.; Leblans, N.I.W.; Woebken, D.; Janssens, I.A.; Sigurdsson, B.D.; Richter, A. Microbial temperature sensitivity and biomass change explain soil carbon loss with warming. *Nat. Clim. Chang.* **2018**, *8*, 885–889. [CrossRef]
30. Nottingham, A.T.; Whitaker, J.; Ostle, N.J.; Bardgett, R.D.; McNamara, N.P.; Fierer, N.; Salinas, N.; Ccahuana, A.J.Q.; Turner, B.L.; Meir, P. Microbial responses to warming enhance soil carbon loss following translocation across a tropical forest elevation gradient. *Ecol. Lett.* **2019**, *22*, 1889–1899. [CrossRef]
31. Cui, P.Y.; Fan, F.L.; Yin, C.; Song, A.L.; Huang, P.R.; Tang, Y.J.; Zhu, P.; Peng, C.; Li, T.Q.; Wakelin, S.A.; et al. Long-term organic and inorganic fertilization alters temperature sensitivity of potential N_2O emissions and associated microbes. *Soil Biol. Biochem.* **2016**, *93*, 131–141. [CrossRef]
32. Jannis, H.; Shurong, L.; Harry, V.; Nicolas, B. Abiotic nitrous oxide production from hydroxylamine in soils and their dependence on soil properties. *Soil Biol. Biochem.* **2015**, *84*, 107–115.
33. Li, J.; Junmin, P.; Elise, P.; Changming, F.; Ming, N. Spatial heterogeneity of temperature sensitivity of soil respiration: A global analysis of field observations. *Soil Biol. Biochem.* **2020**, *141*, 107675. [CrossRef]
34. Hu, H.; Chen, D.; He, J.-Z. Microbial regulation of terrestrial nitrous oxide formation: Understanding the biological pathways for prediction of emission rates. *FEMS Microbiol. Rev.* **2015**, *39*, 729–749. [CrossRef]
35. Hu, H.; Xu, Z.; He, J. Ammonia-Oxidizing Archaea Play a Predominant Role in Acid Soil Nitrification. *Adv. Agron.* **2014**, *125*, 261–302.
36. Liao, B.; Wu, X.; Yu, Y.F.; Luo, S.Y.; Hu, R.G.; Lu, G.A. Effects of mild alternate wetting and drying irrigation and mid-season drainage on CH_4 and N_2O emissions in rice cultivation. *Sci. Total Environ.* **2020**, *698*, 134212. [CrossRef]
37. Liang, K.; Zhong, X.; Huang, N.; Lampayan, R.M.; Liu, Y.; Pan, J.; Peng, B.; Hu, X.; Fu, Y. Nitrogen losses and greenhouse gas emissions under different N and water management in a subtropical double-season rice cropping system. *Sci. Total Environ.* **2017**, *609*, 46–57. [CrossRef]
38. Tirol-Padre, A.; Minamikawa, K.; Tokida, T.; Wassmann, R.; Yagi, K. Site-specific feasibility of alternate wetting and drying as a greenhouse gas mitigation option in irrigated rice fields in Southeast Asia: A synthesis. *Soil Sci. Plant Nutr.* **2018**, *64*, 2–13. [CrossRef]
39. Humphrey, V.; Berg, A.; Ciais, P.; Gentine, P.; Jung, M.; Reichstein, M.; Seneviratne, S.I.; Frankenberg, C. Soil moisture-atmosphere feedback dominates land carbon uptake variability. *Nature* **2021**, *592*, 65–69. [CrossRef]
40. Sylvie, D.; Katell, Q. Analytical pyrolysis as a tool to probe soil organic matter. *J. Anal. Appl. Pyrolysis* **2015**, *111*, 108–120.
41. Li, Y.; Dong, S.; Liu, S.; Zhou, H.; Gao, Q.; Cao, G.; Wang, X.; Su, X.; Zhang, Y.; Tang, L.; et al. Seasonal changes of CO_2, CH_4 and N_2O fluxes in different types of alpine grassland in the Qinghai-Tibetan Plateau of China. *Soil Biol. Biochem.* **2015**, *80*, 306–314. [CrossRef]
42. Chen, H.; Yang, Z.; Chu, R.K.; Tolic, N.; Liang, L.; Graham, D.E.; Wullschleger, S.D.; Gu, B. Molecular Insights into Arctic Soil Organic Matter Degradation under Warming. *Environ. Sci. Technol.* **2018**, *52*, 4555–4564. [CrossRef]
43. Paré, M.C.; Bedard-Haughn, A. Soil organic matter quality influences mineralization and GHG emissions in cryosols: A field-based study of sub- to high Arctic. *Glob. Chang. Biol.* **2013**, *19*, 1126–1140. [CrossRef]
44. Wang, H.H.; Shen, M.X.; Hui, D.F.; Chen, J.; Sun, G.F.; Wang, X.; Lu, C.Y.; Sheng, J.; Chen, L.G.; Luo, Y.Q.; et al. Straw incorporation influences soil organic carbon sequestration, greenhouse gas emission, and crop yields in a Chinese rice (*Oryza sativa* L.)-wheat (*Triticum aestivum* L.) cropping system. *Soil Tillage Res.* **2019**, *195*, 104377. [CrossRef]

45. Pascual, J.; Wust, P.K.; Geppert, A.; Foesel, B.U.; Huber, K.J.; Overmann, J. Novel isolates double the number of chemotrophic species and allow the first description of higher taxa in Acidobacteria subdivision 4. *Syst. Appl. Microbiol.* **2015**, *38*, 534–544. [CrossRef]
46. Xu, P.; Zhou, W.; Jiang, M.D.; Khan, I.; Wu, T.T.; Zhou, M.H.; Zhu, B.; Hu, R.G. Methane emission from rice cultivation regulated by soil hydrothermal condition and available carbon and nitrogen under a rice-wheat rotation system. *Plant Soil* **2022**, 1–12. [CrossRef]
47. Wu, Y.P.; Liu, T.; Peng, Q.; Shaaban, M.; Hu, R.G. Effect of straw returning in winter fallow in Chinese rice fields on greenhouse gas emissions: Evidence from an incubation study. *Soil Res.* **2015**, *53*, 298–305. [CrossRef]
48. Cui, Y.; Meng, J.; Wang, Q.; Weiming, Z.; Weiming, Z.; Wenfu, C. Effects of straw and biochar addition on soil nitrogen, carbon, and super rice yield in cold waterlogged paddy soils of North China. *J. Integr. Agric.* **2017**, *16*, 1064–1074. [CrossRef]
49. Zhao, S.X.; Schmidt, S.; Qin, W.; Li, J.; Li, G.X.; Zhang, W.F. Towards the circular nitrogen economy—A global meta-analysis of composting technologies reveals much potential for mitigating nitrogen losses. *Sci. Total Environ.* **2020**, *704*, 135401. [CrossRef]
50. Awais, S.; Saba, S.; Abdul, R.; Fatima, A.; Muhammad, A.; Sher, M.S.; Taimoor, H.F.; Muhammad, A.; Muhammad, A.M.; Muhammad, M.A.; et al. Effect of animal manure, crop type, climate zone, and soil attributes on greenhouse gas emissions from agricultural soils-A global meta-analysis. *J. Clean. Prod.* **2021**, *278*, 124019.
51. Qu, Z.; Wang, J.; Almoy, T.; Bakken, L.R. Excessive use of nitrogen in Chinese agriculture results in high $N_2O/(N_2O+N_2)$ product ratio of denitrification, primarily due to acidification of the soils. *Glob. Chang. Biol.* **2014**, *20*, 1685–1698. [CrossRef] [PubMed]
52. Shaaban, M.; Wu, Y.P.; Peng, Q.A.; Lin, S.; Mo, Y.L.; Wu, L.; Hu, R.G.; Zhou, W. Effects of dicyandiamide and dolomite application on N_2O emission from an acidic soil. *Environ. Sci. Pollut. Res.* **2016**, *23*, 6334–6342. [CrossRef] [PubMed]
53. Galbally, I.E.; Meyer, C.P.; Wang, Y.P.; Smith, C.J.; Weeks, I.A. Nitrous oxide emissions from a legume pasture and the influences of liming and urine addition. *Agric. Ecosyst. Environ.* **2016**, *136*, 262–272. [CrossRef]
54. Shaaban, M.; Peng, Q.-A.; Hu, R.; Wu, Y.; Lin, S.; Zhao, J. Dolomite application to acidic soils: A promising option for mitigating N_2O emissions. *Environ. Sci. Pollut. Res.* **2015**, *22*, 19961–19970. [CrossRef]
55. Wu, H.; Hao, X.; Xu, P.; Hu, J.; Jiang, M.; Shaaban, M.; Zhao, J.; Wu, Y.; Hu, R. CO_2 and N_2O emissions in response to dolomite application are moisture dependent in an acidic paddy soil. *J. Soils Sediments* **2020**, *20*, 3136–3147. [CrossRef]
56. Shaaban, M.; Wu, Y.P.; Wu, L.; Hu, R.G.; Younas, A.; Nunez-Delgado, A.; Xu, P.; Sun, Z.; Lin, S.; Xu, X.Y.; et al. The Effects of pH Change through Liming on Soil N_2O Emissions. *Processes* **2020**, *8*, 702. [CrossRef]
57. Yang, Y.; Sun, K.; Liu, J.; Chen, Y.L.; Han, L.F. Changes in soil properties and CO_2 emissions after biochar addition: Role of pyrolysis temperature and aging. *Sci. Total Environ.* **2022**, *839*, 156333. [CrossRef]
58. Gasco, G.; Paz-Ferreiro, J.; Cely, P.; Plaza, C.; Mendez, A. Influence of pig manure and its biochar on soil CO_2 emissions and soil enzymes. *Ecol. Eng.* **2016**, *95*, 19–24. [CrossRef]
59. Benavente, I.; Gasco, G.; Plaza, C.; Paz-Ferreiro, J.; Mendez, A. Choice of pyrolysis parameters for urban wastes affects soil enzymes and plant germination in a Mediterranean soil. *Sci. Total Environ.* **2018**, *634*, 1308–1314. [CrossRef]
60. Yu, Z.; Chen, L.; Pan, S.; Li, Y.; Kuzyakov, Y.; Xu, J.; Brookes, P.C.; Luo, Y. Feedstock determines biochar-induced soil priming effects by stimulating the activity of specific microorganisms. *Eur. J. Soil Sci.* **2018**, *69*, 521–534. [CrossRef]
61. Yang, F.; Xu, Z.B.; Huang, Y.D.; Tsang, D.C.W.; Ok, Y.S.; Zhao, L.; Qiu, H.; Xu, X.Y.; Cao, X.D. Stabilization of dissolvable biochar by soil minerals: Release reduction and organo-mineral complexes formation. *J. Hazard. Mater.* **2021**, *412*, 125213. [CrossRef] [PubMed]
62. Yu, M.J.; Su, W.Q.; Parikh, S.J.; Li, Y.; Tang, C.X.; Xu, J.M. Intact and washed biochar caused different patterns of nitrogen transformation and distribution in a flooded paddy soil. *J. Clean. Prod.* **2021**, *293*, 126259. [CrossRef]
63. Wang, L.; Gao, C.C.; Yang, K.; Sheng, Y.Q.; Xu, J.; Zhao, Y.X.; Lou, J.; Sun, R.; Zhu, L.Z. Effects of biochar aging in the soil on its mechanical property and performance for soil CO_2 and N_2O emissions. *Sci. Total Environ.* **2021**, *782*, 146824. [CrossRef] [PubMed]
64. Chen, D.; Wang, C.; Shen, J.L.; Li, Y.; Wu, J.S. Response of CH_4 emissions to straw and biochar applications in double-rice cropping systems: Insights from observations and modeling. *Environ. Pollut.* **2018**, *235*, 95–103. [CrossRef] [PubMed]
65. Wang, C.; Shen, J.L.; Liu, J.Y.; Qin, H.L.; Yuan, Q.; Fan, F.L.; Hu, Y.J.; Wang, J.; Wei, W.X.; Li, Y.; et al. Microbial mechanisms in the reduction of CH_4 emission from double rice cropping system amended by biochar: A four-year study. *Soil Biol. Biochem.* **2019**, *135*, 251–263. [CrossRef]
66. He, L.L.; Shan, J.; Zhao, X.; Wang, S.Q.; Yan, X.Y. Variable responses of nitrification and denitrification in a paddy soil to long-term biochar amendment and short-term biochar addition. *Chemosphere* **2019**, *234*, 558–567. [CrossRef] [PubMed]
67. Lourenco, K.S.; Dimitrov, M.R.; Pijl, A.; Soares, J.R.; Do Carmo, J.B.; van Veen, J.A.; Cantarella, H.; Kuramae, E.E. Dominance of bacterial ammonium oxidizers and fungal denitrifiers in the complex nitrogen cycle pathways related to nitrous oxide emission. *GCB Bioenergy* **2018**, *10*, 645–660. [CrossRef]
68. Mothapo, N.V.; Chen, H.H.; Cubeta, M.A.; Shi, W. Nitrous oxide producing activity of diverse fungi from distinct agroecosystems. *Soil Biol. Biochem.* **2013**, *66*, 94–101. [CrossRef]
69. Ji, C.; Han, Z.Q.; Zheng, F.W.; Wu, S.; Wang, J.Y.; Wang, J.D.; Zhang, H.; Zhang, Y.C.; Liu, S.W.; Li, S.Q.; et al. Biochar reduced soil nitrous oxide emissions through suppressing fungal denitrification and affecting fungal community assembly in a subtropical tea plantation. *Agric. Ecosyst. Environ.* **2022**, *326*, 107784. [CrossRef]
70. Zhang, K.; Wang, Y.; Mao, J.; Chen, B. Effects of biochar nanoparticles on seed germination and seedling growth. *Environ. Pollut.* **2020**, *256*, 113409. [CrossRef]

71. Jan, M.; Josephine, G.; Munoo, P.; Ulf, L.; Juergen, K.; Ondrej, M.; Wolfram, B. Toxicity screening of biochar-mineral composites using germination tests. *Chemosphere* **2018**, *207*, 91–100.
72. Huang, C.; Weiyue, W.; Shizhong, Y.; Muhammad, A.; Yuhui, Q. Role of biochar and Eisenia fetida on metal bioavailability and biochar effects on earthworm fitness. *Environ. Pollut.* **2020**, *263*, 114586. [CrossRef] [PubMed]
73. Malev, O.; Contin, M.; Licen, S.; Barbieri, P.; De Nobili, M. Bioaccumulation of polycyclic aromatic hydrocarbons and survival of earthworms (Eisenia andrei) exposed to biochar amended soils. *Environ. Sci. Pollut. Res.* **2016**, *23*, 3491–3502. [CrossRef] [PubMed]
74. Li, L.J.; You, M.Y.; Shi, H.A.; Ding, X.L.; Qiao, Y.F.; Han, X.Z. Soil CO_2 emissions from a cultivated Mollisol: Effects of organic amendments, soil temperature, and moisture. *Eur. J. Soil Biol.* **2013**, *55*, 83–90. [CrossRef]
75. Wang, C.; Ma, X.F.; Shen, J.L.; Chen, D.; Zheng, L.; Ge, T.D.; Li, Y.; Wu, J.S. Reduction in net greenhouse gas emissions through a combination of pig manure and reduced inorganic fertilizer application in a double-rice cropping system: Three-year results. *Agric. Ecosyst. Environ.* **2022**, *326*, 107799. [CrossRef]
76. Li, B.Z.; Song, H.; Cao, W.C.; Wang, Y.J.; Chen, J.S.; Guo, J.H. Responses of soil organic carbon stock to animal manure application: A new global synthesis integrating the impacts of agricultural managements and environmental conditions. *Glob. Chang. Biol.* **2021**, *27*, 5356–5367. [CrossRef]
77. Tian, H.Q.; Lu, C.Q.; Ciais, P.; Michalak, A.M.; Canadell, J.G.; Saikawa, E.; Huntzinger, D.N.; Gurney, K.R.; Sitch, S.; Zhang, B.W.; et al. The terrestrial biosphere as a net source of greenhouse gases to the atmosphere. *Nature* **2016**, *531*, 225–228. [CrossRef]
78. Yuan, J.; Yuan, Y.; Zhu, Y.; Cao, L. Effects of different fertilizers on methane emissions and methanogenic community structures in paddy rhizosphere soil. *Sci. Total Environ.* **2018**, *627*, 770–781. [CrossRef]
79. Li, Y.J.; Zheng, Q.; Yang, R.; Zhuang, S.; Lin, W.; Li, Y.Z. Evaluating microbial role in reducing N_2O emission by dual isotopocule mapping following substitution of inorganic fertilizer for organic fertilizer. *J. Clean. Prod.* **2021**, *326*, 129442. [CrossRef]
80. Guo, L.J.; Zhang, L.; Liu, L.; Sheng, F.; Cao, C.G.; Li, C.F. Effects of long-term no tillage and straw return on greenhouse gas emissions and crop yields from a rice-wheat system in central China. *Agric. Ecosyst. Environ.* **2021**, *322*, 107650. [CrossRef]
81. Wu, X.H.; Wang, W.; Xie, K.J.; Yin, C.M.; Hou, H.J.; Xie, X.L. Combined effects of straw and water management on CH_4 emissions from rice fields. *J. Environ. Manag.* **2019**, *231*, 1257–1262. [CrossRef] [PubMed]
82. Wang, X.D.; He, C.; Cheng, H.Y.; Liu, B.Y.; Li, S.S.; Wang, Q.; Liu, Y.; Zhao, X.; Zhang, H.L. Responses of greenhouse gas emissions to residue returning in China's croplands and influential factors: A meta-analysis. *J. Environ. Manag.* **2021**, *289*, 112486. [CrossRef] [PubMed]
83. Li, H.; Dai, M.W.; Dai, S.L.; Dong, X.J. Current status and environment impact of direct straw return in China's cropland—A review. *Ecotoxicol. Environ. Saf.* **2018**, *159*, 293–300. [CrossRef] [PubMed]
84. Liu, C.Y.; Wang, K.; Meng, S.X.; Zheng, X.H.; Zhou, Z.X.; Han, S.H.; Chen, D.L.; Yang, Z.P. Effects of irrigation, fertilization and crop straw management on nitrous oxide and nitric oxide emissions from a wheat-maize rotation field in northern China. *Agric. Ecosyst. Environ.* **2011**, *140*, 226–233. [CrossRef]
85. Xu, C.; Han, X.; Ru, S.H.; Cardenas, L.; Rees, R.M.; Wu, D.; Wu, W.L.; Meng, F.Q. Crop straw incorporation interacts with N fertilizer on N_2O emissions in an intensively cropped farmland. *Geoderma* **2019**, *341*, 129–137. [CrossRef]
86. Chen, Z.X.; Tu, X.S.; Meng, H.; Chen, C.; Chen, Y.J.; Elrys, A.S.; Cheng, Y.; Zhang, J.B.; Cai, Z.C. Microbial process-oriented understanding of stimulation of soil N_2O emission following the input of organic materials. *Environ. Pollut.* **2021**, *284*, 117176. [CrossRef]
87. Zhang, Z.S.; Guo, L.J.; Liu, T.Q.; Li, C.F.; Cao, C.G. Effects of tillage practices and straw returning methods on greenhouse gas emissions and net ecosystem economic budget in rice wheat cropping systems in central China. *Atmos. Environ.* **2015**, *122*, 636–644. [CrossRef]
88. Zhu, H.J.; Tao, J.; Yan, X.M.; Zhou, B.J.; Mwangi, J.K. Short-Term Effects of Straw Application on Carbon Recycle in a Rice-Rapeseed Rotation System. *Aerosol Air Qual. Res.* **2016**, *16*, 3358–3363. [CrossRef]
89. Jiang, C.M.; Yu, W.T.; Ma, Q.; Xu, Y.G.; Zou, H. Alleviating global warming potential by soil carbon sequestration: A multi-level straw incorporation experiment from a maize cropping system in Northeast China. *Soil Tillage Res.* **2017**, *170*, 77–84. [CrossRef]
90. Liang, S.; Zhang, H. Different Responses of Greenhouse Gas Emissions to Straw Application at Different Seasons in Northeast China. *Sains Malays.* **2019**, *48*, 1347–1355. [CrossRef]
91. Zhang, H.; Liang, S.; Wang, Y.H.; Liu, S.W.; Sun, H.D. Greenhouse gas emissions of rice straw return varies with return depth and soil type in paddy systems of Northeast China. *Arch. Agron. Soil Sci.* **2021**, *67*, 1591–1602. [CrossRef]
92. Li, S.H.; Guo, L.J.; Cao, C.G.; Li, C.F. Effects of straw returning levels on carbon footprint and net ecosystem economic benefits from rice-wheat rotation in central China. *Environ. Sci. Pollut. Res.* **2021**, *28*, 5742–5754. [CrossRef]
93. Wu, G.; Ling, J.; Xu, Y.P.; Zhao, D.Q.; Liu, Z.X.; Wen, Y.; Zhou, S.L. Effects of soil warming and straw return on soil organic matter and greenhouse gas fluxes in winter wheat seasons in the North China Plain. *J. Clean. Prod.* **2022**, *356*, 131810. [CrossRef]
94. Alvarez, A.L.; Weyers, S.L.; Goemann, H.M.; Peyton, B.M.; Gardner, R.D. Microalgae, soil and plants: A critical review of microalgae as renewable resources for agriculture. *Algal Res.* **2021**, *54*, 102200. [CrossRef]
95. de Siqueira, C.J.; Lucia, C.M.; Peixoto, A.P.; Roberto, C.P.; Rodrigues, D.A.I.; Jose, R.V. Microalgae biofilm in soil: Greenhouse gas emissions, ammonia volatilization and plant growth. *Sci. Total Environ.* **2017**, *574*, 1640–1648.
96. Bharti, A.; Velmourougane, K.; Prasanna, R. Phototrophic biofilms: Diversity, ecology and applications. *J. Appl. Phycol.* **2017**, *29*, 2729–2744. [CrossRef]

97. Marks, E.A.N.; Minon, J.; Pascual, A.; Montero, O.; Navas, L.M.; Rad, C. Application of a microalgal slurry to soil stimulates heterotrophic activity and promotes bacterial growth. *Sci. Total Environ.* **2017**, *605*, 610–617. [CrossRef]
98. Knoche, K.L.; Aoyama, E.; Hasan, K.; Minteer, S.D. Role of Nitrogenase and Ferredoxin in the Mechanism of Bioelectrocatalytic Nitrogen Fixation by the Cyanobacteria Anabaena variabilis SA-1 Mutant Immobilized on Indium Tin Oxide (ITO) Electrodes. *Electrochim. Acta* **2017**, *232*, 396–403. [CrossRef]
99. Mallappa, M.; Amrita, K.; Kunal, R.; Siddarthan, V.; Radha, P.; Balasubramanian, R.; Firoz, H.; Lata, N.; Yashbir, S.S.; Awadhesh, B.R.; et al. Beneficial cyanobacteria and eubacteria synergistically enhance bioavailability of soil nutrients and yield of okra. *Heliyon* **2016**, *2*, e00066.
100. Renuka, N.; Prasanna, R.; Sood, A.; Ahluwalia, A.S.; Bansal, R.; Babu, S.; Singh, R.; Shivay, Y.S.; Nain, L. Exploring the efficacy of wastewater-grown microalgal biomass as a biofertilizer for wheat. *Environ. Sci. Pollut. Res.* **2016**, *23*, 6608–6620. [CrossRef]
101. Ali, M.A.; Sattar, M.A.; Islam, M.N.; Inubushi, K. Integrated effects of organic, inorganic and biological amendments on methane emission, soil quality and rice productivity in irrigated paddy ecosystem of Bangladesh: Field study of two consecutive rice growing seasons. *Plant Soil* **2014**, *378*, 239–252. [CrossRef]
102. Prasanna, R.; Adak, A.; Verma, S.; Bidyarani, N.; Babu, S.; Pal, M.; Shivay, Y.S.; Nain, L. Cyanobacterial inoculation in rice grown under flooded and SRI modes of cultivation elicits differential effects on plant growth and nutrient dynamics. *Ecol. Eng.* **2015**, *84*, 532–541. [CrossRef]
103. Shrestha, R.C.; Ghazaryan, L.; Poodiack, B.; Zorin, B.; Gross, A.; Gillor, O.; Khozin-Goldberg, I.; Gelfand, I. The effects of microalgae-based fertilization of wheat on yield, soil microbiome and nitrogen oxides emissions. *Sci. Total Environ.* **2022**, *806*, 151320. [CrossRef]
104. Zhang, S.P.; Wang, L.; Wei, W.; Hu, J.J.; Mei, S.H.; Zhao, Q.Y.; Tsang, Y.F. Enhanced roles of biochar and organic fertilizer in microalgae for soil carbon sink. *Biodegradation* **2018**, *29*, 313–321. [CrossRef]
105. Hu, J.J.; Guo, H.C.; Xue, Y.Y.; Gao, M.T.; Zhang, S.P.; Tsang, Y.F.; Li, J.X.; Wang, Y.N.; Wang, L. Using a mixture of microalgae, biochar, and organic manure to increase the capacity of soil to act as carbon sink. *J. Soils Sediments* **2019**, *19*, 3718–3727. [CrossRef]
106. Bhardwaj, D.; Ansari, M.W.; Sahoo, R.K.; Tuteja, N. Biofertilizers function as key player in sustainable agriculture by improving soil fertility, plant tolerance and crop productivity. *Microb. Cell Fact.* **2014**, *13*, 66. [CrossRef]
107. Yen, H.W.; Ho, S.H.; Chen, C.Y.; Chang, J.S. CO_2, NO_x and SO_x removal from flue gas via microalgae cultivation: A critical review. *Biotechnol. J.* **2015**, *10*, 829–839. [CrossRef]
108. Zeraatkar, A.K.; Ahmadzadeh, H.; Talebi, A.F.; Moheimani, N.R.; McHenry, M.P. Potential use of algae for heavy metal bioremediation, a critical review. *J. Environ. Manag.* **2016**, *181*, 817–831. [CrossRef]

processes

MDPI

Review

A Review on the Hydrothermal Treatment of Food Waste: Processing and Applications

Chuanbin Wang [1], Zhi Wang [1], Xutong Wang [1], Ning Li [1], Junyu Tao [2], Wandong Zheng [1], Beibei Yan [1], Xiaoqiang Cui [1,*], Zhanjun Cheng [1,*] and Guanyi Chen [2,3,4]

1 Tianjin Key Laboratory of Biomass Waste Utilization, School of Environmental Science and Engineering, Tianjin University, Tianjin 300072, China
2 School of Mechanical Engineering, Tianjin University of Commerce, Tianjin 300134, China
3 School of Science, Tibet University, Lhasa 850012, China
4 Tianjin Engineering Research Center for Organic Wastes Safe Disposal and Energy Utilization, School of Environmental Science and Engineering, Tianjin University, Tianjin 300072, China
* Correspondence: cuixiaoqiang@tju.edu.cn (X.C.); zjcheng@tju.edu.cn (Z.C.)

Abstract: The amount of food waste is increasing with the development of society and the increase in population; the rough treatment of food waste could result in a serious environmental crisis and waste of resources. Hydrothermal treatment is a promising scheme to achieve the harmless treatment and utilization of food waste. Although there are many studies on the hydrothermal treatment of food waste, there is still a lack of systematic summary and comprehensive analysis of the relevant literature. In this review, we provide an in-depth analysis of the specific impact mechanisms of hydrothermal conditions on the gaseous, solid, and liquid products. Meanwhile, the hydrothermal conversion mechanisms of food waste components are systematically sorted out. The review also discusses the potential application areas for the derived products from the hydrothermal treatment of food waste. Finally, the main challenges and future research directions are proposed to improve the development of the hydrothermal treatment of food waste.

Keywords: food waste; hydrothermal treatment; influence condition; applications

1. Introduction

With the improvement in the economic level and the improvement of people's living standards, the catering industry has developed extremely rapidly, and the amount of food waste generated shows an increasing trend. Food waste accounts for a large proportion of solid waste, probably ranging from 10% to 33% [1]. According to the food sustainability index 2017 report, the national ranking of food waste production per capita is: Australia (361 kg) > United States (287 kg) > Sweden (200 kg) > Russia (56 kg) > China (44 kg) [2]. Although China's per capita production of food waste is relatively small, the total amount of food waste produced in China is huge due to its large population.

Food waste is the wasted or lost part of the food supply chain [3]. In a broad sense, food waste refers to organic waste generated in the production, sales, processing, and consumption of food, including solid food residues, catering waste water, waste edible oils, etc. [4]. Besides, food waste is mainly composed of starch, dietary fiber, fat, protein, and other organic matter, which has the characteristics of high moisture content and high organic matter [5]. However, food waste can easily induce many serious environmental problems without proper treatment. For example, food waste will produce malodorous gas and sewage in the process of storage and transportation, which will bring great environmental hazards if these pollutants are directly discharged into the surrounding environment.

In the past, nearly 80% of food waste could not be properly treated and recycled, which would cause great damage to the environment and waste resources [6]. Organic carbon and other nutrients in food waste could be converted into energy materials and value-added products through suitable treatment processes [7,8]. At present, food waste treatment technology mainly includes landfill, incineration, anaerobic digestion, composting, and hydrothermal treatment technology. The advantages and disadvantages of different treatment methods are shown in Table 1. The hydrothermal method is particularly suitable for the treatment of wet organic waste [9]. The hydrothermal treatment of food waste is a new resource utilization method with great advantages, which could help reduce the amount of food waste in landfills, protect the environment, and reduce greenhouse gases at the same time [10]. Besides, the economic cost is low, because there is no need for drying [11], and the product has a high utilization value. Therefore, the hydrothermal treatment of food waste is an efficient way of resource utilization, which has great value in reduction, harmlessness, and resource utilization for food waste.

Table 1. Advantages and disadvantages of different food waste treatment methods.

Treatment Process	Advantages	Disadvantages
Landfill	Low cost, simple technology, more used in developing countries	Pollution of groundwater, occupying a large amount of land, and no resource recovery
Incineration	High degree of reduction, suitable for handling hazardous or toxic garbage	High cost, long capital recovery cycle, and low economic efficiency
Anaerobic digestion	High degree of automation, diversified products, high economic value	Need to screen suitable microorganisms, complex technology, discontinuous cycle, and difficult-to-treat biogas residue
Compost	High technology maturity, low cost	Low product value, environmental pollution, long cycle
Hydrothermal treatment	The product has high energy utilization value, simple process, low cost, and short cycle	Hydrothermal treatment products need to be further optimized

From 2010 to 2021 (according to ISI Web of Science™), publications related to the hydrothermal treatment of food waste have been steadily increasing (Figure 1). Research related to the hydrothermal treatment of food waste progressed slowly from 2010 to 2016 and accelerated after 2017. In addition, research on the hydrothermal carbonization of food waste, which was relatively scarce before 2015, became a mainstream research process in recent years, probably because hydrothermal carbonization products had a wide range of applications. For hydrothermal liquefaction as well as hydrothermal gasification, there had been a trend of increasing research year by year, but the growth rate was relatively slow.

Considering that few papers fully describe the relevant information on the hydrothermal treatment of food waste, this review focuses on the processing and the products of the hydrothermal treatment of food waste. All the relevant literature is collected and organized into three categories: hydrothermal carbonization, hydrothermal liquefaction, and hydrothermal gasification for analysis. Firstly, the classification of the hydrothermal process is introduced (Section 2), and then the specific influence mechanisms of hydrothermal parameters on the three-phase products are discussed (Section 3). Afterward, the hydrothermal conversion mechanisms of different biomass components were explored in detail (Section 4). As an important part, the potential applications of the value-added solid, liquid, and gas products derived from hydrothermal treatment of food waste are particularly summarized and discussed (Section 5). Finally, the challenges and development directions are proposed for the industrialization of the hydrothermal treatment of food waste (Section 6). The schematic illustration of this paper was shown in Figure 2.

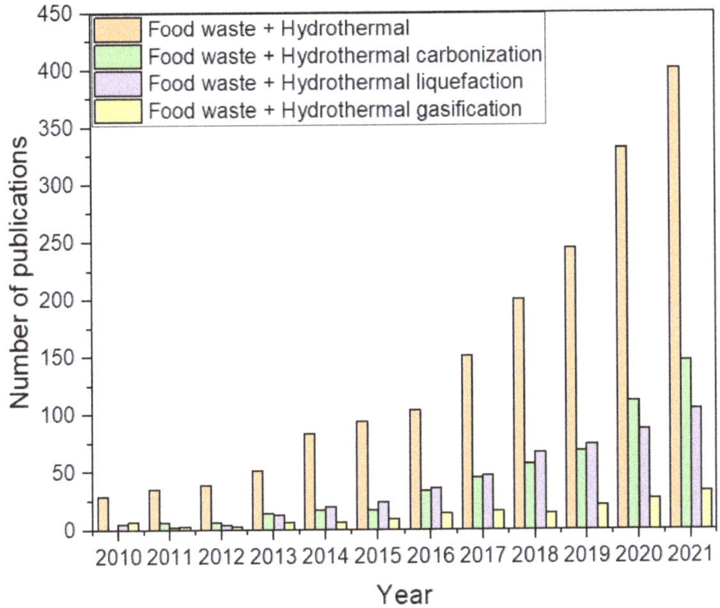

Figure 1. Science Citation Indexed publications on the hydrothermal treatments of food waste in Web of Science.

Figure 2. Schematic illustration of this paper.

2. Hydrothermal Process

Hydrothermal treatment of biomass is a very efficient thermochemical conversion process which can transform biomass into valuable biofuels or value-added products [12,13]. According to the pressure and temperature range, the hydrothermal process is divided into hydrothermal carbonization, hydrothermal liquefaction, and hydrothermal gasification.

At present, there are many studies on hydrothermal liquefaction and hydrothermal carbonization and relatively few studies on hydrothermal gasification. Hydrothermal gasification requires higher temperature and pressure, and higher requirements for equipment safety. Table 2 listed the conditions of the three hydrothermal methods.

Hydrothermal carbonization mainly converts biomass into carbon materials (hydrochar) in the temperature range of 180–260 °C under 1–8 MPa [12]. Food residue is hydrothermally carbonized to produce fuel particles, and the prepared solid particles have good combustion properties and can be used as household fuels [14]. Food waste hydrochar is used as other value-added products [15]. For example, food waste could be prepared by hydrothermal carbonization to prepare N-doped carbon dots, which had high selectivity to Fe^{3+} fluorescent probes and could be used as bioimaging materials [16].

Hydrothermal liquefaction occurs generally at 260–374 °C under 10–22.1 MPa [17]. The main product is high-energy bio-oil and other products. Bio-oil is a mixture of substances with different molecular weights such as alcohol, sugar, and furan [9]. The bio-oil produced by the hydrothermal liquefaction of food waste had good quality. The heating value of bio-oil was as high as 34.79 $MJ \cdot kg^{-1}$, which was expected to become an alternative fuel for petroleum [18]. In addition, food waste is rich in organic matter, which can be hydrothermally liquefied to extract useful products. For example, the extraction rate of pectin was as high as 11.63% by microwave-assisted hydrothermal treatment, and pectin could be used as a thickener and had important commercial value [19].

The temperature range of hydrothermal gasification is above 400 °C, and the energy consumption is very high. The main target product of this stage is synthesis gas, which mainly includes hydrogen and methane. Supercritical water gasification of food residues could produce hydrogen synthesis gas. The supercritical water gasification treatment method is a reliable treatment method for the resource utilization of food waste [20].

Table 2. Classification of different hydrothermal treatment processes for food waste [21].

Hydrothermal Process	Temperature Range (°C)	Pressure Range	Target Product
Hydrothermal carbonization	180–260	1–8 MPa	Hydrochar
Hydrothermal liquefaction	260–374	10–22.1 MPa	Oil
Hydrothermal gasification	>374	>22.1 MPa	Syngas

3. Factors Affecting the Hydrothermal Process of Food Waste

3.1. Temperature

Temperature is the most important influencing factor in the hydrothermal process, and increasing temperature generally contributes to the reduction in food waste. An increase in temperature will lead to a decrease in the production of hydrochar, but it will enhance the combustion performance of hydrochar. Temperature also plays a key role in the production of bio-oil. The elevated temperature will lead to an increase in the production of bio-oil, which has a certain impact on the physical and chemical properties of bio-oil, such as viscosity. In addition, temperature also affects the carbohydrates and other components in the solution. It is necessary to control the appropriate temperature to adjust the content of related components in the oil and liquid phase products. The production of gas in the hydrothermal process is also greatly affected by temperature. Generally, the increase in temperature can promote the production of clean energy gases such as hydrogen and methane.

3.1.1. The Influence of Temperature on Solid Products

Temperature has an important influence on the production of hydrochar. Figure 3 showed the influence of temperature on hydrothermal products. The production of hydrochar decreased from 45% to 35% when the temperature increased from 225 to 250 °C due to the intensification of decarboxylation and dehydration reactions [22,23]. In addition, temperature had an important influence on the physical and chemical properties of hydrochar. The increase in temperature caused the HHV of hydrochar to increase from 15 MJ/kg to 31 MJ/kg. At the same time, high temperature helped to improve the combustion performance and density of hydrochar [24,25]. Conversely, the increase in temperature lead to a decrease in the atomic ratio of O/C and (O&N)/C, indicating that the polarity of hydrochar decreased with increasing temperature [26].

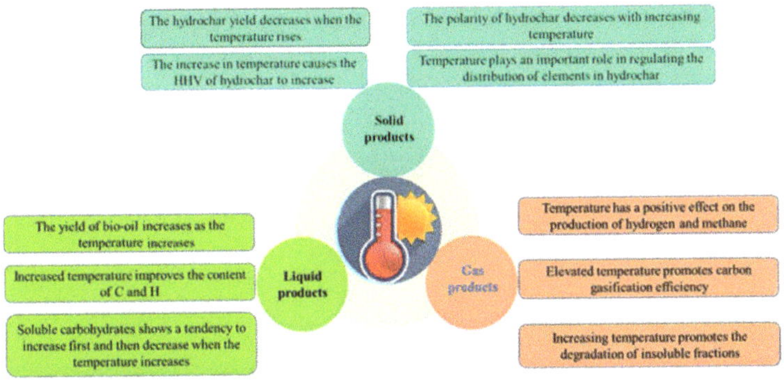

Figure 3. The influence of temperature on hydrothermal products.

Temperature plays an important role in regulating the distribution of elements in hydrochar [27]. The increase in temperature facilitated the removal of ammonia and enriched the nitrogen content of aromatic heterocyclic compounds in the hydrochar [28]. In addition, the increased temperature reduced the production of hydrochar [29], but the content of carbon [16], volatile substances [30], calcium, magnesium and other minerals in the hydrochar elevated [31]. Furthermore, the removal effect of N, S, and Cl was more obvious when the temperature increased from 180 to 260 °C [32]. Especially, with the increase in temperature, the content of pyridine nitrogen and graphitic nitrogen increased, and the use time of carbon dots was shorter, while the content of amino nitrogen and pyrrolic nitrogen decreased [33]. At the same time, temperature was one of the important factors that determined the fate of phosphorus [34].

3.1.2. The Influence of Temperature on Liquid Products

The main components of bio-oil produced by the hydrothermal liquefaction of kitchen waste are C_6–C_{22} compounds [35]. The yield of bio-oil increased significantly as the temperature increased [36–38], but the energy conversion efficiency declined with the increase in temperature [39]. Therefore, it is necessary to determine the appropriate temperature to ensure economic feasibility and energy conversion efficiency [40].

For biofuel preparation, the increased temperature improved the content of C and H and decreased the viscosity. At the same time, the increase in temperature caused the decomposition of organic acids in food waste, leading to an elevated pH of the solution [41]. In addition, the total organic carbon of the liquid also showed a downward trend with the increase in the temperature, indicating that the increase in temperature had a certain promotion effect on the decomposition of water-soluble products [42]. Besides, the soluble carbohydrates showed a tendency to increase first and then decrease when the hydrothermal temperature increased from 100 to 200 °C [43].

3.1.3. The Influence of Temperature on Gas Products

Biogas is generated from food waste through hydrothermal gasification and hydrothermal pretreatment combined with fermentation or other processes. Temperature had a positive effect on the production of hydrogen and methane [5]. What is more, elevated temperature promoted carbon gasification efficiency when studying the kinetics of supercritical water gasification of food waste [4]. The hydrothermal pretreatment temperature of food waste had a certain influence on the production of biogas (hydrogen and methane) produced by fermentation. Increasing temperature promoted the degradation of insoluble fractions, which was conducive to the generation of biogas [44–46].

3.2. *Pyrolysis Time*

The reaction time is an important factor in the hydrothermal process [47], but the influence degree of the reaction time often needs to be determined in conjunction with other reaction conditions. Generally, under other conditions being fixed, increasing the reaction time helped to increase biogas production, but inhibited the production of hydrochar [48]. The effect of time on hydrothermal products was displayed in Figure 4.

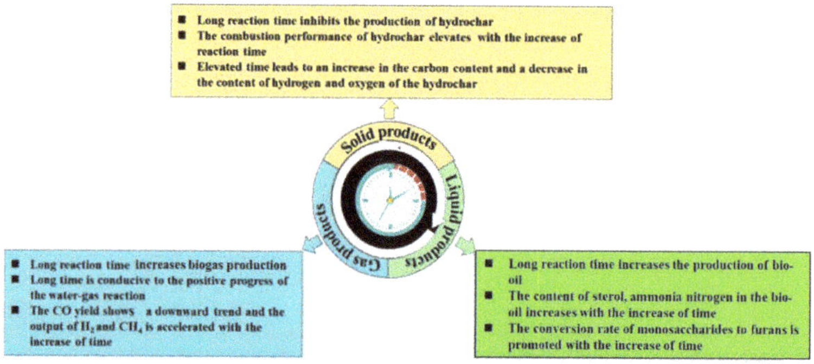

Figure 4. The influence of reaction time on hydrothermal products.

3.2.1. The Effect of Time on Solid Products

The extension of time promoted the condensation reaction of degradation products and increased the conversion rate of hydrochar [49], which reduced the yield of hydrochar [50,51]. In addition, the combustion performance of hydrochar improved and the energy consumption was elevated with the increase in reaction time [52,53]. The elevated time leaded to an increase in the carbon content and a decrease in the content of hydrogen and oxygen in the hydrochar [29]. In addition, the desulfurization of hydrochar was promoted as the prolonged hydrothermal time [53], and the chlorine content in the hydrochar was inversely proportional to the residence time [54]. Besides, residence time had a certain effect on the shape and crystal structure of the hydrochar [55].

3.2.2. The Effect of Time on Liquid Products

Too short a time is not conducive to the hydrothermal reaction of food waste and affects the quality of bio-oil, and excessive reaction time would increase production costs [41]. Under other conditions unchanged, the biotransformation rate and bio-oil output would increase slightly with the time increased [36,38]. However, there was a limit to prolonging the reaction time, increasing the reaction time was not conducive production of bio-oil if the reaction time exceeded the limit [8].

In addition, hydrothermal treatment of olive oil waste was investigated, and the total sterol content in the bio-oil increased significantly when the time increased from 15 to 90 min [56]. As the reaction time was prolonged, the content of ammonia nitrogen in liquid products increased [34], and the conversion rate of monosaccharides to furans was promoted [3].

3.2.3. The Effect of Time on Gas Products

Longer reaction times help to increase the conversion of food waste and increase the gas-phase yield [38]. The prolongation of the reaction time was conducive to the positive progress of the water-gas reaction, resulting in a downward trend in CO yield and an upward trend in hydrogen production [57]. However, the output of CO_2 and CH_4 accelerated with the increase in time due to the hydrolysis, decarboxylation, and dehydration reactions [20,58].

3.3. Catalyst

The catalyst could reduce the activation energy of the hydrothermal reaction, which is an important influencing factor. At present, the relevant catalysts for the hydrothermal process of food waste mainly include the following types: alkaline catalysts, acid catalysts, metal-related catalysts, and biocatalysts. Each type of catalyst has different effects on the hydrothermal process of food waste. The hydrothermal catalysis process requires a lower operating environment, and the energy density and the yield of biofuels increase significantly [59]. Therefore, it is necessary to select a suitable catalyst according to the target product to achieve the purpose of optimal treatment of food waste.

3.3.1. Alkaline Catalysts

Different catalysts are suitable for different feedstocks in the hydrothermal process. Alkaline catalysts have certain requirements on the acidity of lipid feedstocks [60], and the use of basic catalysts in the hydrothermal liquefaction process may increase the hydrolysis of macromolecules, reduce the formation of char, and stabilize the production of bio-oil [61].

Alkaline catalysts such as K_3PO_4, K_2CO_3, Na_3PO_4, and Na_2CO_3 have a good effect on the production of biodiesel and affect the reaction rate in the transesterification reaction. The activity of the catalysts from high to low was $K_2CO_3 > K_3PO_4 > Na_3PO_4 > Na_2CO_3$. The effect of potassium salt was relatively good, but its recovery rate and stability were not good [62].

As the usage of K_2CO_3 increases, the production of hydrochar decreased, and the production of bio-oil and gas increased, indicating that hydrochar could be decomposed to produce tar compounds [42,60]. In addition, the presence of alkaline catalysts promoted the decomposition of short-chain compounds into simple gases. NaOH could inhibit the formation of char and tar, resulting in a higher gas yield. As the concentration of the K_2CO_3 catalyst increased, the yield of H_2 and CH_4 increased, while the yield of CO decreased [20]. Rattana Muangrat explored the influence of alkaline catalysts on the hydrogen produced, the results showed that the H_2 production rate was in the order of NaOH > KOH > $Ca(OH)_2$ > K_2CO_3 > Na_2CO_3 > $NaHCO_3$ [63]. The catalyst could promote the production of H_2 from kitchen waste by regulating the water-gas shift reaction while suppressing the production of bio-oil and water char. In addition, NaOH could absorb part of CO_2 and make the water-gas reaction move positively, thereby increasing the H_2 production rate [63]. Thus, sodium hydroxide is an efficient catalyst for the production of hydrogen by hydrothermal gasification.

3.3.2. Acid Catalysts

The acid-base catalyst is used to convert biomass into bio-oil through a hydrothermal reaction, the quality of biofuel is improved under the action of the acid catalyst [64]. Acid catalysts has the advantages in the conversion of cellulose raw materials for bio-oil production [38]. In addition, acidic conditions might enhance dehydration reactions to produce biocrude with lower H/C ratios [65].

Phosphoric acid contributed to the production of volatile fatty acids during the hydrothermal treatment of food waste [66]. In addition, sulfuric acid concentration had an important effect on the efficiency of food waste conversion to 5-hydroxymethylfurfural. The product yield increased first and then decreased when the catalyst concentration was 0.5 M. However, the product yield increased first and then decreased when the catalyst concentration was 1.0 M [67].

3.3.3. Metal-Related Catalysts

The bio-oil performance improved after catalysis using a ceria catalyst [68]. The homogeneous catalyst Na_2CO_3 or heterogeneous catalyst CeZrOx was used to promote the reaction. The results showed that CeZrOx had a better effect, and the prepared bio-oil had a higher heating value (HHV) and energy recovery [10]. In addition, the concentration of the catalyst has a certain influence on the hydrothermal process of food waste. Sodium methoxide with a concentration of 0.9 wt.% could maximize the yield of methyl esters, and increasing the catalyst concentration would not increase the yield [69]. Furthermore, multicomponent metal oxides had a better catalytic effect than single metal oxides, which might be due to the synergistic effect of acid sites and base sites on the surface of mixed metal catalysts [70]. For example, hydrogen production was the best under the condition of adding two catalysts (Na_2CO_3 and Ni/SiO_2), so there was a synergistic effect between the two catalysts [71].

3.3.4. Biocatalysts

Biocatalysts also have good catalytic performance [72]. Clamshell-derived catalyst is used to catalyze the hydrothermal liquefaction of algae. The bio-oil production could reach 39.6 wt.% when the catalyst loading rate was 0.6 wt.%. In addition, the bio-oil production rate dropped to 37.1 wt.% when the catalyst loading rate increased to 1 wt.% [73]. In addition, fish bone carbon was used to catalyze waste cooking palm oil to produce biodiesel. When the catalyst loading was 2.5 wt.%, the transesterification reaction yield reached up to 98%, because the higher density of alkaline active sites enhanced the catalytic effect [74]. Furthermore, waste coffee grounds were used to prepare nitrogen-doped solid biochar. The biochar was used as a catalyst for the hydrothermal reaction of glucose. The results showed that the nitrogen-rich biochar had an excellent catalytic performance on glucose and helped to quickly convert glucose into fructose, and the reaction activity was equivalent to that of a basic catalyst [75].

3.4. Raw Material

The composition of food waste raw materials is the most fundamental factor affecting the hydrothermal process. The biochemical composition of food waste has a significant effect on both the oil yield and various product compositions and properties [8]. The catalyst, temperature, and time affect the reaction process based on the raw material composition. The influence of the chemical composition of cellulose, hemicellulose, and lignin on the properties of hydrochar was explored, and the lignin hydrochar had a large yield, rough surface, and more benzene rings. In addition, the hydrochar produced by hemicellulose had higher microsphere density and oxygen-containing groups [76]. Furthermore, the effects of carbohydrates, proteins, and lipids on the combustion and pyrolysis characteristics of food waste hydrochar were studied, and carbohydrates helped to enhance the thermal stability of hydrochar. Hydrothermal pretreatment could facilitate the fermentation process and increase the production of volatile fatty acids from vegetable protein products, while there was no significant increase in the production of volatile fatty acids from animal protein products [77]. In addition, the samples with higher starch content produced higher calorific values, and the crude biological product rich in starch had the highest HHV (42 MJ/kg) [78]. However, protein had little effect on hydrochar combustion characteristics [79]. What is more, increased lipid content promoted the combustion reaction and loss of more volatiles, bio-oil yields for meat and cheese were as high as 60% and 75%, respectively [80]. Lipid-based bio-oil had the highest yield and protein-based bio-oil had the lowest yield [81].

Table 3 showed the hydrochar yield of food waste and its calorific value. The yield of hydrochar from food waste ranged from 43% to 66%. Among these, the high content of cellulose and hemicellulose in rabbit food and sweet corn were easily hydrothermally converted under high-temperature conditions, which lead to a lower hydrochar yield. The presence of cellulose and hemicellulose facilitated the dehydration reaction during hydrothermal carbonization [82]. In addition, the higher yield of hydrochar of a mixture of food waste and yard waste was due to the fact that yard waste was rich in lignin, which was more difficult to convert into bio-oil or biogas than cellulose and hemicellulose [83]. The energy content of the hydrochar varied from 10 to 35 MJ/kg depending on the raw material. The high calorific value of hydrochar in restaurant food waste might be attributed to the high content of proteins and lipids [84–87], which could be hydrothermally converted into small molecules with high calorific value. Therefore, the composition of the raw material is crucial to the yield and heating value of the derived hydrochar.

Table 3. Hydrochar yield of food waste and its calorific value.

Feedstock	Temperature (°C)	Hydrochar Yield (%)	Heating Value (MJ/kg)	Reference
Dog food	234–295	55.0	26.0	[88]
Rabbit food	250	43.8	29.1	[89]
Restaurant food waste	225–275	45.0	33.6	[27]
Sweet corn	250	50.0	11.0	[90]
Grape pomace	175–275	46.5	24.3–28.3	[91]
Food waste from a student's hostel mess	200	48.5	30.0	[54]
Food waste and yard waste	220	59.8	27.6	[83]

Moisture content and food waste concentration have important effects on the production of value-added products from hydrothermal treatment. The increase in food concentration was not conducive to the production of hydrogen, the output of CO, CO_2, CH_4, and C_2–C_4 all increased as the food waste concentration increases [57].

Further, the binary mixtures with the synergistic effect of hydrothermal liquefaction on crude bioproduct yield are as follows: protein and cellulose, protein and xylose, cellulose and lignin, and xylose and lignin. On the contrary, soybean oil has an antagonistic effect with lignin [92]. In addition, mixed hydrothermal treatment of food waste with other substances is also an important resource utilization method [93]. Hydrochar prepared by mixed hydrothermal carbonization of kitchen waste and woody biomass had good combustion performance and was suitable for use as solid fuel [94–96]. In addition, the hydrochar prepared by mixing food waste and coal according to the ratio of 1:1 had the best calorific value (31.4 MJ/Kg) [25]. Currently, food waste hydrochar could be used in the related fields of daily life and production in conjunction with other fuels [24]. Furthermore, the mixed hydrothermal carbonization of food waste and packaging materials was investigated, and the concentration of raw materials affected the carbon distribution in the product due to solubilization [27]. Although the existence of plastic was not conducive to the stability of hydrochar, the lower content of plastic promoted the production of gas and improved the reaction efficiency [97].

4. Hydrothermal Reaction Pathways

Food waste is mainly composed of protein, lipids, carbohydrates, cellulose, hemicellulose, and lignin. The hydrothermal conversion mechanism refers to the conversion process of raw materials reacting under hydrothermal conditions to produce the target products. The composition of food waste is complex, and the reaction mechanisms involve a variety of parallel chemical reactions. The specific details of the specific reaction path are unclear and difficult to study. Therefore, the transformation mechanisms of the hydrothermal process of food waste components needs to be explored in detail.

Figure 5 showed a schematic diagram of the hydrothermal process of different biomass components. The conversion process mainly involved five steps: hydrolysis, dehydration, decarboxylation, condensation, polymerization, and aromatization [6]. Lipid, protein, cellulose, hemicellulose, and lignin were hydrolyzed into oligomers or monomers at relatively low temperatures. Dehydration and decarboxylation as well as the polymerization process also occurred during the hydrolysis process [6]. In addition, the increase in temperature caused the decarboxylation reaction and Maillard reaction to proceed, decomposing the monomer material [98]. The hydrolysis of carbohydrates was relatively easy, and the products were simple acids and hydrocarbons with low carbon content. The products of protein hydrothermal reaction were amines, cyclic amides, organic acids, and other substances.

Besides, the hydrothermal process of lipids would undergo the halogenation reaction to produce alkane halides.

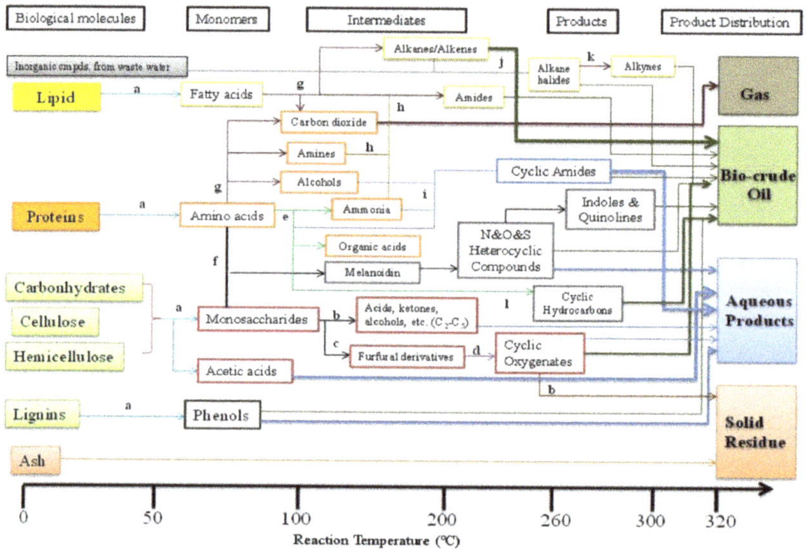

Figure 5. Hydrothermal reaction path of raw material components.: (**a**) hydrolysis; (**b**) decomposition; (**c**) dehydration; (**d**) polymerization; (**e**) deamination; (**f**) Maillard reaction; (**g**) decarboxylation; (**h**) aminolysis; (**i**) cyclization; (**j**) halogenations; (**k**) dehydrohalogenation; (**l**) condensation + pyrolysis. Reprinted from [98] with permission from Elsevier.

5. Applications

In recent years, food waste is an untapped resource with great potential for application. Food waste can produce value-added products through hydrothermal treatment, and hydrochar, gas, and oil products can be widely used in all aspects of society.

5.1. Solid Product

Figure 6 showed the applications of the solid product obtained by hydrothermal treatment of food waste. Firstly, the hydrothermal carbonization of food waste could produce multifunctional application materials and biofuels [99], indicating that hydrothermal carbonization was an effective way of resource utilization of food waste [100]. The generated food waste hydrochar had a higher heating value (higher heating value = ~30 MJ/kg), which was comparable to lignite [54]. The solid fuel produced by hydrothermal carbonization of food waste and other waste could replace traditional fuels [101,102]. In addition, food waste was very abundant and had broad application prospects. The solid fuel prepared by the hydrothermal carbonization of food waste and yard waste could replace up to approximately 11% of global coal consumption [83]. Furthermore, food waste undergoed hydrothermal carbonization to synthesize solid acid and realized the conversion of food waste into biofuel [64].

Carbon dots have good compatibility and fluorescence characteristics, which have important uses in the fields of biomedicine, environmental monitoring, and photocatalysis [103]. Food waste hydrochar could be used to prepare multi-color carbon dots to achieve high-value utilization of food waste [104]. In addition, the nitrogen-doped photoluminescent carbon nanodots generated from fish scales were rich in nitrogen and had rich heteroatom groups, which had good application value [105]. Besides, the carbon dots prepared from pork ribs could be used for rapid detection of dimethoate, and the preparation process had the advantages of low cost and simple operation [103].

Figure 6. Applications of the solid product obtained by hydrothermal treatment of food waste.

In addition, food waste could be converted into other high-value-added products through hydrothermal carbonization. Formate is a good battery material and hydrogen storage carrier. The food waste was hydrothermally carbonized to prepare formate, and the yield of food residues converted into formate was up to 78% [106]. In addition, eggshells, pomace, and clam shells as a carbon source could be hydrothermally synthesized hydroxyapatite [55,107,108]. The hydroxyapatite extracted from pomelo peel had the best performance, and its shape was similar to the structure of crystalline hydroxyapatite in human bones [55]. Furthermore, bread waste was converted into high-quality graphene flakes by hydrothermal synthesis at a cost of only 3 $\$/m^2$, which was far lower than the current market price (56,000 $\$/m^2$) [109].

At the same time, hydrochar has certain adsorption and catalytic capability [110]. Food waste hydrochar is a potential high-efficiency adsorbent for water pollutants [111], and the specific surface area and pore volume are important factors that affect the adsorption capacity [110]. Food waste hydrochar had a good removal effect on La^{3+}, and the maximum adsorption capacity was up to 108 $mg \cdot g^{-1}$ [112]. In addition, hydrothermal carbonization could recover nutrients from food waste. The hydrochar and liquid products produced by food waste could replace about 0.96% and 2.30% of nitrogen and phosphorus-based fertilizers, respectively [34].

Food waste can also be converted into materials for industrial applications through hydrothermal treatment. For example, vegetable waste was converted into acetic acid by hydrothermal method, and then the calcium acetate produced was used as a new green deicing agent [113]. In addition, the carbon-based Fe_3O_4 nanocomposite prepared by the hydrothermal reaction of pomelo peel was a good fruit magnetic solid-phase extraction material [114]. Furthermore, the hydrochar produced from food waste had a similar structure to pure sugars and could be used for application [75,110,115].

5.2. Liquid Product

The applications of liquid material obtained by hydrothermal treatment of food waste were displayed in Figure 7. High-quality bio-oil was produced from food waste through a hydrothermal treatment, which provided an effective way for the resource utilization of food waste [116,117]. The hydrothermal liquefaction of waste edible oil can efficiently produce green biodiesel [72].

In addition, calcined bone could act as a catalyst to promote the transesterification reaction, significantly increasing the yield of biodiesel (up to 98%) [74]. Hydrothermal pretreatment combined with anaerobic fermentation was a high-value utilization method of food waste resources [118]. The chili post-harvest residue was subjected to surfactant hydrothermal pretreatment and fermentation treatment to produce bioethanol, which realized the utilization of waste resources. In addition, hydrothermal pretreatment did not require any additional treatment, and the treatment process was simple.

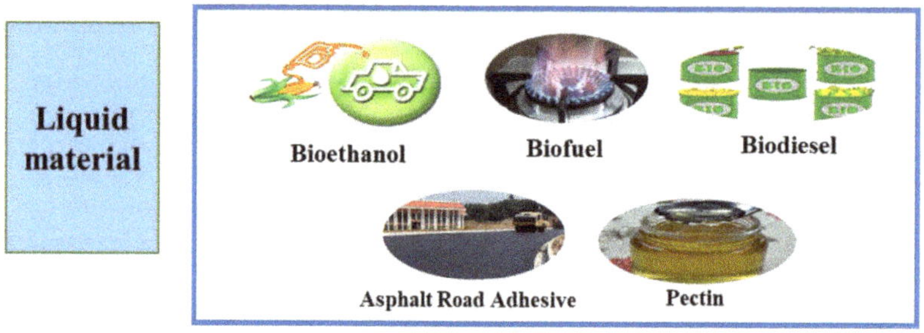

Figure 7. Applications of liquid material obtained by hydrothermal treatment of food waste.

Food waste (banana peel, watermelon peel, lemon peel, and tomato waste) was hydrothermally liquefied to produce bio-oil. The addition of a catalyst (K_2CO_3) could significantly improve the quality and calorific value of bio-oil [38,40,48]. The heating value of bio-oil was as high as 38.059 kJ/kg [68]. The quality of bio-oil was upgraded to bio-diesel through the distillation and esterification processes, indicating that bio-oil might become a substitute for fossil fuels in the future [119]. Besides, enzyme pretreatment helped with the production of bio-crude oil in the hydrothermal treatment process of food waste [120].

In addition, the hydrothermal treatment of food waste can produce a variety of other high-value-added products. For instance, 1 ton of watermelon rind could produce 1.52 kg of 5-hydroxymethyl furfural through hydrothermal liquefaction, which had certain economic feasibility and market application potential [121]. Besides, the microwave-assisted hydrothermal treatment of mango peel could achieve efficient extraction of pectin, and the extraction rate was as high as 11.63%. In general, pectin could be used as a thickening agent and had important commercial value [19]. Furthermore, the product extracted from food waste through hydrothermal liquefaction could be used as a green binder for asphalt roads, which helped to reduce the use of crude oil [122].

5.3. Gas Product

Food waste can generate natural gas (hydrogen and methane) through hydrothermal treatment. The preparation process is relatively complex and is divided into a single preparation process and a coupled preparation process. The single preparation process includes hydrothermal gasification and supercritical water gasification. The coupling preparation process includes hydrothermal reaction combined with aerobic fermentation, hydrothermal liquefaction anaerobic digestion, hydrothermal pretreatment combined with supercritical hydro thermalization, hydrothermal treatment combined with steam gasification technology, and hydrothermal pretreatment combined with composting. Hydrothermal treatment contributes to the reduction in food waste and the disposal of resources.

The cellulose is hydrothermally gasified, and the selective generation of hydrogen could be achieved at a relatively low hydrothermal temperature [71]. Hydrothermal pretreatment of food waste can break down cellulose into simple substances such as glucose and can increase the ratio of protein to carbohydrate, which is beneficial for subsequent fermentation and other processing [123].

Hydrothermal pretreatment combined with anaerobic fermentation could produce biogas [37,124,125]. The hydrothermal process has a greater destructive effect on food waste and promotes the production of hydrogen [126]. The hydrogen production obtained at 200 °C reached the maximum value of 81.27 mL/gVS [15]. Besides, the methane yield was as high as 511.6 mL/g, and the energy conversion efficiency was as high as 78.6% [43]. Furthermore, hydrothermal pretreatment combined with the aerobic fermentation process could control malodorous gases [127]. For instance, hydrogen production from hydrothermal pretreatment combined with anaerobic fermentation of coffee waste was seven times higher than that by the single fermentation process [128]. Thus, the combined use of hydrothermal pretreatment and anaerobic fermentation has good application prospects.

Hydrothermal pretreatment could reduce energy consumption and reduce costs when food waste undergoes hydrothermal pretreatment combined with supercritical water gasification to produce hydrogen [58]. In addition, hydrochar steam gasification is a potentially efficient process for hydrogen production [42].

6. Challenges and Development Directions of Hydrothermal Treatment of Food Waste

1. Economic analysis of the hydrothermal process of food waste is scarce, so it is necessary to analyze the economic efficiency of hydrothermal treatment to promote industrial applications.
2. At present, the main research is intermittent experimental research, and there is almost no research on continuous hydrothermal treatment of food waste. In addition, there is no relevant continuous hydrothermal equipments on the market, so it is necessary to increase the research and development of relative equipment.
3. The influence of reaction conditions has been discussed a lot regarding temperature and time, but the specific impact mechanisms of the catalyst are unclear. At the same time, there are few studies on influencing factors such as the aqueous phase cycle, pressure, and heating rate. Thus, the detailed study of hydrothermal experimental parameters should be strengthened in the future.
4. In terms of the application of hydrothermal products, there is less research on the utilization of all components of oil, hydrochar and gas products, so it is necessary to strengthen research in this area to improve the high-value utilization of food waste.
5. Most of the current reaction mechanisms are derived from model compounds, which is very different from the actual hydrothermal mechanism of food waste. Therefore, it is necessary to study the reaction mechanism of real food waste and explore the interaction mechanism between different components.

7. Conclusions

Food waste is a potential application resource. However, it is not suitable for incineration or pyrolysis due to the high moisture content of food waste. The hydrothermal process, which does not require the drying of food waste, could be used to prepare value-added products and has attracted widespread attention. The hydrothermal treatment process is mainly divided into three types: hydrothermal carbonization, hydrothermal liquefaction, and hydrothermal gasification. Temperature, time, feedstock composition, and catalysts have important effects on the product. Temperature and catalysts are the main external factors on the yields and properties of hydrothermal products. The interconversion of the solid-liquid-gas products could be achieved by changing the temperature and catalyst, so the suitable hydrothermal processing temperature and catalyst should be selected purposefully according to the target products. In addition, the composition of food waste, an important intrinsic factor in the hydrothermal product, should be concerned first for the generation of target products. The solid, liquid, and gas products derived from the hydrothermal treatment of food waste show good potential in energy production, environmental remediation, and new material preparation, while challenges remain. Hydrochar has good prospects for application in environmental remediation and energy storage, but the high production cost limits its practical application. As for the bio-oil with a certain amount of oxygen, a cost-effective strategy for bio-oil upgradation should be developed. Moreover, considering the complex components of the gas products, the separation and purification of the gases urgently need to be solved. Finally, the exploration of the hydrothermal treatment process and the design of hydrothermal equipments should be strengthened to promote the resource utilization of food waste in the future.

Author Contributions: Methodology, Formal analysis, Data Curation, Writing—Original draft preparation, C.W.; Resource, Writing—Review and Editing, Z.W.; Writing—Review and Editing, X.W.; Conceptualization, Writing-Review and Editing, N.L.; Resource, Writing—Review and Editing, J.T.; Writing—Review and Editing, Funding, W.Z.; Supervision, B.Y.; Resource, Data Curation, Writing—Review and Editing, X.C.; Conceptualization, Writing—Review and Editing, Z.C.; Writing—Review and Editing, G.C. All authors have read and agreed to the published version of the manuscript.

Funding: This research was funded by Ministry of Science and Technology of the People's Republic of China (2021YFE0102500).

Institutional Review Board Statement: Not applicable.

Informed Consent Statement: Informed consent was obtained from all subjects involved in the study.

Conflicts of Interest: The authors declare no conflict of interest.

References

1. Ajay, C.M.; Mohan, S.; Dinesha, P. Decentralized energy from portable biogas digesters using domestic kitchen waste: A review. *Waste Manag.* **2021**, *125*, 10–26. [CrossRef] [PubMed]
2. Srivastava, N.; Srivastava, M.; Abd_Allah, E.F.; Singh, R.; Hashem, A.; Gupta, V.K. Biohydrogen production using kitchen waste as the potential substrate: A sustainable approach. *Chemosphere* **2021**, *271*, 129537. [CrossRef] [PubMed]
3. Li, F.; Liu, L.; An, Y.; He, W.; Themelis, N.J.; Li, G. Hydrothermal liquefaction of three kinds of starches into reducing sugars. *J. Clean. Prod.* **2016**, *112*, 1049–1054. [CrossRef]
4. Chen, J.; Fan, Y.; Jiaqiang, E.; Cao, W.; Zhang, F.; Gong, J.; Liu, G.; Xu, W. Effects analysis on the gasification kinetic characteristics of food waste in supercritical water. *Fuel* **2019**, *241*, 94–104. [CrossRef]
5. Su, H.; Hantoko, D.; Yan, M.; Cai, Y.; Kanchanatip, E.; Liu, J.; Zhou, X.; Zhang, S. Evaluation of catalytic subcritical water gasification of food waste for hydrogen production: Effect of process conditions and different types of catalyst loading. *Int. J. Hydrog. Energy* **2019**, *44*, 21451–21463. [CrossRef]
6. Zhou, Y.; Engler, N.; Nelles, M. Symbiotic relationship between hydrothermal carbonization technology and anaerobic digestion for food waste in China. *Bioresour. Technol.* **2018**, *260*, 404–412. [CrossRef] [PubMed]
7. Pham, T.P.; Kaushik, R.; Parshetti, G.K.; Mahmood, R.; Balasubramanian, R. Food waste-to-energy conversion technologies: Current status and future directions. *Waste Manag.* **2015**, *38*, 399–408. [CrossRef] [PubMed]
8. Aierzhati, A.; Stablein, M.J.; Wu, N.E.; Kuo, C.T.; Si, B.; Kang, X.; Zhang, Y. Experimental and model enhancement of food waste hydrothermal liquefaction with combined effects of biochemical composition and reaction conditions. *Bioresour. Technol.* **2019**, *284*, 139–147. [CrossRef]
9. Déniel, M.; Haarlemmer, G.; Roubaud, A.; Weiss-Hortala, E.; Fages, J. Energy valorisation of food processing residues and model compounds by hydrothermal liquefaction. *Renew. Sustain. Energy Rev.* **2016**, *54*, 1632–1652. [CrossRef]
10. Maag, A.; Paulsen, A.; Amundsen, T.; Yelvington, P.; Tompsett, G.; Timko, M. Catalytic Hydrothermal Liquefaction of Food Waste Using CeZrOx. *Energies* **2018**, *11*, 564. [CrossRef]
11. Fernandez-Sanroman, A.; Lama, G.; Pazos, M.; Rosales, E.; Sanroman, M.A. Bridging the gap to hydrochar production and its application into frameworks of bioenergy, environmental and biocatalysis areas. *Bioresour. Technol.* **2021**, *320 Pt B*, 124399. [CrossRef]
12. Tekin, K.; Karagöz, S.; Bektaş, S. A review of hydrothermal biomass processing. *Renew. Sustain. Energy Rev.* **2014**, *40*, 673–687. [CrossRef]
13. Zhang, W.; Qiu, X.; Wang, C.; Zhong, L.; Fu, F.; Zhu, J.; Zhang, Z.; Qin, Y.; Yang, D.; Xu, C.C. Lignin derived carbon materials: Current status and future trends. *Carbon Res.* **2022**, *1*, 14. [CrossRef]
14. Zhai, Y.; Wang, T.; Zhu, Y.; Peng, C.; Wang, B.; Li, X.; Li, C.; Zeng, G. Production of fuel pellets via hydrothermal carbonization of food waste using molasses as a binder. *Waste Manag.* **2018**, *77*, 185–194. [CrossRef]
15. Parshetti, G.K.; Chowdhury, S.; Balasubramanian, R. Hydrothermal conversion of urban food waste to chars for removal of textile dyes from contaminated waters. *Bioresour. Technol.* **2014**, *161*, 310–319. [CrossRef] [PubMed]
16. Ahn, J.; Song, Y.; Kwon, J.E.; Woo, J.; Kim, H. Characterization of food waste-driven carbon dot focusing on chemical structural, electron relaxation behavior and Fe^{3+} selective sensing. *Data Brief* **2019**, *25*, 104038. [CrossRef]
17. Shen, Y. A review on hydrothermal carbonization of biomass and plastic wastes to energy products. *Biomass Bioenergy* **2020**, *134*, 105479. [CrossRef]
18. Chen, W.-H.; Lin, Y.-Y.; Liu, H.-C.; Chen, T.-C.; Hung, C.-H.; Chen, C.-H.; Ong, H.C. A comprehensive analysis of food waste derived liquefaction bio-oil properties for industrial application. *Appl. Energy* **2019**, *237*, 283–291. [CrossRef]
19. Matharu, A.S.; Houghton, J.A.; Lucas-Torres, C.; Moreno, A. Acid-free microwave-assisted hydrothermal extraction of pectin and porous cellulose from mango peel waste—Towards a zero waste mango biorefinery. *Green Chem.* **2016**, *18*, 5280–5287. [CrossRef]
20. Nanda, S.; Isen, J.; Dalai, A.K.; Kozinski, J.A. Gasification of fruit wastes and agro-food residues in supercritical water. *Energy Convers. Manag.* **2016**, *110*, 296–306. [CrossRef]

21. Kambo, H.S.; Dutta, A. A comparative review of biochar and hydrochar in terms of production, physico-chemical properties and applications. *Renew. Sustain. Energy Rev.* **2015**, *45*, 359–378. [CrossRef]
22. Akarsu, K.; Duman, G.; Yilmazer, A.; Keskin, T.; Azbar, N.; Yanik, J. Sustainable valorization of food wastes into solid fuel by hydrothermal carbonization. *Bioresour. Technol.* **2019**, *292*, 121959. [CrossRef] [PubMed]
23. Li, C.; Li, J.; Pan, L.; Zhu, X.; Xie, S.; Yu, G.; Wang, Y.; Pan, X.; Zhu, G.; Angelidaki, I. Treatment of digestate residues for energy recovery and biochar production: From lab to pilot-scale verification. *J. Clean. Prod.* **2020**, *265*, 121852. [CrossRef]
24. Saqib, N.U.; Baroutian, S.; Sarmah, A.K. Physicochemical, structural and combustion characterization of food waste hydrochar obtained by hydrothermal carbonization. *Bioresour. Technol.* **2018**, *266*, 357–363. [CrossRef]
25. Ul Saqib, N.; Sarmah, A.K.; Baroutian, S. Effect of temperature on the fuel properties of food waste and coal blend treated under co-hydrothermal carbonization. *Waste Manag.* **2019**, *89*, 236–246. [CrossRef]
26. Fu, M.-M.; Mo, C.-H.; Li, H.; Zhang, Y.-N.; Huang, W.-X.; Wong, M.H. Comparison of physicochemical properties of biochars and hydrochars produced from food wastes. *J. Clean. Prod.* **2019**, *236*, 117637. [CrossRef]
27. Li, L.; Diederick, R.; Flora, J.R.; Berge, N.D. Hydrothermal carbonization of food waste and associated packaging materials for energy source generation. *Waste Manag.* **2013**, *33*, 2478–2492. [CrossRef]
28. Wang, T.; Zhai, Y.; Zhu, Y.; Peng, C.; Xu, B.; Wang, T.; Li, C.; Zeng, G. Influence of temperature on nitrogen fate during hydrothermal carbonization of food waste. *Bioresour. Technol.* **2018**, *247*, 182–189. [CrossRef]
29. Chen, X.; Ma, X.; Peng, X.; Lin, Y.; Yao, Z. Conversion of sweet potato waste to solid fuel via hydrothermal carbonization. *Bioresour. Technol.* **2018**, *249*, 900–907. [CrossRef]
30. Theppitak, S.; Hungwe, D.; Ding, L.; Xin, D.; Yu, G.; Yoshikawa, K. Comparison on solid biofuel production from wet and dry carbonization processes of food wastes. *Appl. Energy* **2020**, *272*, 115264. [CrossRef]
31. Erdogan, E.; Atila, B.; Mumme, J.; Reza, M.T.; Toptas, A.; Elibol, M.; Yanik, J. Characterization of products from hydrothermal carbonization of orange pomace including anaerobic digestibility of process liquor. *Bioresour. Technol.* **2015**, *196*, 35–42. [CrossRef] [PubMed]
32. Wang, T.; Zhai, Y.; Zhu, Y.; Gan, X.; Zheng, L.; Peng, C.; Wang, B.; Li, C.; Zeng, G. Evaluation of the clean characteristics and combustion behavior of hydrochar derived from food waste towards solid biofuel production. *Bioresour. Technol.* **2018**, *266*, 275–283. [CrossRef] [PubMed]
33. Ahn, J.; Song, Y.; Kwon, J.E.; Lee, S.H.; Park, K.S.; Kim, S.; Woo, J.; Kim, H. Food waste-driven N-doped carbon dots: Applications for Fe^{3+} sensing and cell imaging. *Mater. Sci. Eng. C Mater. Biol. Appl.* **2019**, *102*, 106–112. [CrossRef] [PubMed]
34. Idowu, I.; Li, L.; Flora, J.R.V.; Pellechia, P.J.; Darko, S.A.; Ro, K.S.; Berge, N.D. Hydrothermal carbonization of food waste for nutrient recovery and reuse. *Waste Manag.* **2017**, *69*, 480–491. [CrossRef]
35. Katakojwala, R.; Kopperi, H.; Kumar, S.; Venkata Mohan, S. Hydrothermal liquefaction of biogenic municipal solid waste under reduced H_2 atmosphere in biorefinery format. *Bioresour. Technol.* **2020**, *310*, 123369. [CrossRef]
36. Demirkaya, E.; Dal, O.; Yüksel, A. Liquefaction of waste hazelnut shell by using sub- and supercritical solvents as a reaction medium. *J. Supercrit. Fluids* **2019**, *150*, 11–20. [CrossRef]
37. Posmanik, R.; Labatut, R.A.; Kim, A.H.; Usack, J.G.; Tester, J.W.; Angenent, L.T. Coupling hydrothermal liquefaction and anaerobic digestion for energy valorization from model biomass feedstocks. *Bioresour. Technol.* **2017**, *233*, 134–143. [CrossRef]
38. Zhang, Y.; Minaret, J.; Yuan, Z.; Dutta, A.; Xu, C. Mild Hydrothermal liquefaction of high water content agricultural residue for bio-crude oil production: A parametric study. *Energies* **2018**, *11*, 3129. [CrossRef]
39. Mahmood, R.; Parshetti, G.K.; Balasubramanian, R. Energy, exergy and techno-economic analyses of hydrothermal oxidation of food waste to produce hydro-char and bio-oil. *Energy* **2016**, *102*, 187–198. [CrossRef]
40. Chen, W.-H.; Lin, Y.-Y.; Liu, H.-C.; Baroutian, S. Optimization of food waste hydrothermal liquefaction by a two-step process in association with a double analysis. *Energy* **2020**, *199*, 117438. [CrossRef]
41. Wang, L.; Chi, Y.; Shu, D.; Weiss-Hortala, E.; Nzihou, A.; Choi, S. Experimental studies of hydrothermal liquefaction of kitchen waste with H^+, OH^- and Fe^{3+} additives for bio-oil upgrading. *Waste Manag. Res.* **2021**, *39*, 165–173. [CrossRef] [PubMed]
42. Duman, G.; Akarsu, K.; Yilmazer, A.; Keskin Gundogdu, T.; Azbar, N.; Yanik, J. Sustainable hydrogen production options from food wastes. *Int. J. Hydrogen Energy* **2018**, *43*, 10595–10604. [CrossRef]
43. Ding, L.; Cheng, J.; Qiao, D.; Yue, L.; Li, Y.Y.; Zhou, J.; Cen, K. Investigating hydrothermal pretreatment of food waste for two-stage fermentative hydrogen and methane co-production. *Bioresour. Technol.* **2017**, *241*, 491–499. [CrossRef] [PubMed]
44. Zieminski, K.; Romanowska, I.; Kowalska-Wentel, M.; Cyran, M. Effects of hydrothermal pretreatment of sugar beet pulp for methane production. *Bioresour. Technol.* **2014**, *166*, 187–193. [CrossRef]
45. Li, M.; Xia, T.; Zhu, C.; Xi, B.; Jia, X.; Wei, Z.; Zhu, J. Effect of short-time hydrothermal pretreatment of kitchen waste on biohydrogen production: Fluorescence spectroscopy coupled with parallel factor analysis. *Bioresour. Technol.* **2014**, *172*, 382–390. [CrossRef]
46. Cheng, J.; Yue, L.; Hua, J.; Dong, H.; Zhou, J.; Li, Y.Y. Hydrothermal alkali pretreatment contributes to fermentative methane production of a typical lipid from food waste through co-production of hydrogen with methane. *Bioresour. Technol.* **2020**, *306*, 123164. [CrossRef]
47. Han, D.; Yeom, K.; Park, S.; Cho, O.; Baek, Y. A Study on the Manufacture of Bio-SRF from the Food Waste by Hydrothermal Carbonization (HTC) Process. *Trans. Korean Hydrog. New Energy Soc.* **2017**, *28*, 426–432. [CrossRef]

48. Zhang, B.; Chen, J.; He, Z.; Chen, H.; Kandasamy, S. Hydrothermal liquefaction of fresh lemon-peel: Parameter optimisation and product chemistry. *Renew. Energy* **2019**, *143*, 512–519. [CrossRef]
49. Zhao, K.; Li, Y.; Zhou, Y.; Guo, W.; Jiang, H.; Xu, Q. Characterization of hydrothermal carbonization products (hydrochars and spent liquor) and their biomethane production performance. *Bioresour. Technol.* **2018**, *267*, 9–16. [CrossRef]
50. Cantero-Tubilla, B.; Cantero, D.A.; Martinez, C.M.; Tester, J.W.; Walker, L.P.; Posmanik, R. Characterization of the solid products from hydrothermal liquefaction of waste feedstocks from food and agricultural industries. *J. Supercrit. Fluids* **2018**, *133*, 665–673. [CrossRef]
51. Zhou, Y.; Engler, N.; Li, Y.; Nelles, M. The influence of hydrothermal operation on the surface properties of kitchen waste-derived hydrochar: Biogas upgrading. *J. Clean. Prod.* **2020**, *259*, 121020. [CrossRef]
52. Gallifuoco, A.; Taglieri, L.; Papa, A.A. Hydrothermal carbonization of waste biomass to fuel: A novel technique for analyzing experimental data. *Renew. Energy* **2020**, *149*, 1254–1260. [CrossRef]
53. Papa, A.A.; Taglieri, L.; Gallifuoco, A. Hydrothermal carbonization of waste biomass: An experimental comparison between process layouts. *Waste Manag.* **2020**, *114*, 72–79. [CrossRef] [PubMed]
54. Gupta, D.; Mahajani, S.M.; Garg, A. Investigation on hydrochar and macromolecules recovery opportunities from food waste after hydrothermal carbonization. *Sci. Total Environ.* **2020**, *749*, 142294. [CrossRef] [PubMed]
55. Wu, S.-C.; Tsou, H.-K.; Hsu, H.-C.; Hsu, S.-K.; Liou, S.-P.; Ho, W.-F. A hydrothermal synthesis of eggshell and fruit waste extract to produce nanosized hydroxyapatite. *Ceram. Int.* **2013**, *39*, 8183–8188. [CrossRef]
56. Lama-Munoz, A.; Rodriguez-Gutierrez, G.; Rubio-Senent, F.; Gomez-Carretero, A.; Fernandez-Bolanos, J. New hydrothermal treatment of alperujo enhances the content of bioactive minor components in crude pomace olive oil. *J. Agric. Food Chem.* **2011**, *59*, 1115–1123. [CrossRef] [PubMed]
57. Muangrat, R.; Onwudili, J.A.; Williams, P.T. Alkali-promoted hydrothermal gasification of biomass food processing waste: A parametric study. *Int. J. Hydrogen Energy* **2010**, *35*, 7405–7415. [CrossRef]
58. Liu, J.; Wang, D.; Yu, C.; Jiang, J.; Guo, M.; Hantoko, D.; Yan, M. A two-step process for energy-efficient conversion of food waste via supercritical water gasification: Process design, products analysis, and electricity evaluation. *Sci. Total Environ.* **2021**, *752*, 142331. [CrossRef]
59. Mackintosh, A.F.; Shin, T.; Yang, H.; Choe, K. Hydrothermal polymerization catalytic process effect of various organic wastes on reaction time, yield, and temperature. *Processes* **2020**, *8*, 303. [CrossRef]
60. Yin, K.; Li, L.; Giannis, A.; Weerachanchai, P.; Ng, B.J.H.; Wang, J.-Y. High-quality fuel from food waste-investigation of a stepwise process from the perspective of technology development. *Environ. Technol.* **2017**, *38*, 1735–1741. [CrossRef]
61. Xu, C.; Lad, N. Production of heavy oils with high caloric values by direct liquefaction of woody biomass in sub/near-critical water. *Energy Fuels* **2008**, *22*, 635–642. [CrossRef]
62. Malins, K. The potential of K_3PO_4, K_2CO_3, Na_3PO_4 and Na_2CO_3 as reusable alkaline catalysts for practical application in biodiesel production. *Fuel Process. Technol.* **2018**, *179*, 302–312. [CrossRef]
63. Muangrat, R.; Onwudili, J.A.; Williams, P.T. Influence of alkali catalysts on the production of hydrogen-rich gas from the hydrothermal gasification of food processing waste. *Appl. Catal. B* **2010**, *100*, 440–449. [CrossRef]
64. Zhang, L.; Tian, L.; Xi, G.; Sun, R.; Zhao, X. Catalytic valorization of expired fructan-rich food into the biofuel 5-ethoxymethylfurfural via a restaurant food waste-derived carbonaceous solid acid. *Waste Biomass Valorization* **2019**, *11*, 6223–6233. [CrossRef]
65. Posmanik, R.; Martinez, C.M.; Cantero-Tubilla, B.; Cantero, D.A.; Sills, D.L.; Cocero, M.J.; Tester, J.W. Acid and alkali catalyzed hydrothermal liquefaction of dairy manure digestate and food waste. *ACS Sustain. Chem. Eng.* **2017**, *6*, 2724–2732. [CrossRef]
66. Shen, D.; Wang, K.; Yin, J.; Chen, T.; Yu, X. Effect of phosphoric acid as a catalyst on the hydrothermal pretreatment and acidogenic fermentation of food waste. *Waste Manag.* **2016**, *51*, 65–71. [CrossRef]
67. Kim, H.S.; Park, M.-R.; Jeon, Y.J.; Kim, S.-K.; Hong, Y.-K.; Jeong, G.-T. Valorization of chitosan as food waste of aquatic organisms into 5-hydroxymethylfurfural by sulfamic acid-catalyzed conversion process. *Energy Technol.* **2018**, *6*, 1747–1754. [CrossRef]
68. Puangubol, S.; Utistham, T.; Wetwatana, U. Production of bio-oil by hydrothermal pyrolysis of food waste over ceria catalyst. *Curr. Opin. Biotechnol.* **2011**, *22*, S49. [CrossRef]
69. Wang, C.; Xie, S.; Zhong, M. Effect of hydrothermal pretreatment on kitchen waste for biodiesel production using alkaline catalyst. *Waste Biomass Valorization* **2016**, *8*, 369–377. [CrossRef]
70. Cheng, F.; Tompsett, G.A.; Murphy, C.M.; Maag, A.R.; Carabillo, N.; Bailey, M.; Hemingway, J.J.; Romo, C.I.; Paulsen, A.D.; Yelvington, P.E.; et al. Synergistic effects of inexpensive mixed metal oxides for catalytic hydrothermal liquefaction of food wastes. *ACS Sustain. Chem. Eng.* **2020**, *8*, 6877–6886. [CrossRef]
71. Ishida, Y.; Kumabe, K.; Hata, K.; Tanifuji, K.; Hasegawa, T.; Kitagawa, K.; Isu, N.; Funahashi, Y.; Asai, T. Selective hydrogen generation from real biomass through hydrothermal reaction at relatively low temperatures. *Biomass Bioenergy* **2009**, *33*, 8–13. [CrossRef]
72. Tran, T.T.; Kaiprommarat, S.; Kongparakul, S.; Reubroycharoen, P.; Guan, G.; Nguyen, M.H.; Samart, C. Green biodiesel production from waste cooking oil using an environmentally benign acid catalyst. *Waste Manag.* **2016**, *52*, 367–374. [CrossRef] [PubMed]

73. Arun, J.; Gopinath, K.P.; SundarRajan, P.; Malolan, R.; Adithya, S.; Sai Jayaraman, R.; Srinivaasan Ajay, P. Hydrothermal liquefaction of Scenedesmus obliquus using a novel catalyst derived from clam shells: Solid residue as catalyst for hydrogen production. *Bioresour. Technol.* **2020**, *310*, 123443. [CrossRef] [PubMed]
74. Chinglenthoiba, C.; Das, A.; Vandana, S. Enhanced biodiesel production from waste cooking palm oil, with NaOH-loaded Calcined fish bones as the catalyst. *Environ. Sci. Pollut. Res. Int.* **2020**, *27*, 15925–15930. [CrossRef]
75. Chen, S.S.; Yu, I.K.M.; Cho, D.-W.; Song, H.; Tsang, D.C.W.; Tessonnier, J.-P.; Ok, Y.S.; Poon, C.S. Selective glucose isomerization to fructose via a nitrogen-doped solid base catalyst derived from spent coffee grounds. *ACS Sustain. Chem. Eng.* **2018**, *6*, 16113–16120. [CrossRef]
76. Xiao, K.; Liu, H.; Li, Y.; Yi, L.; Zhang, X.; Hu, H.; Yao, H. Correlations between hydrochar properties and chemical constitution of orange peel waste during hydrothermal carbonization. *Bioresour. Technol.* **2018**, *265*, 432–436. [CrossRef]
77. Shen, D.; Yin, J.; Yu, X.; Wang, M.; Long, Y.; Shentu, J.; Chen, T. Acidogenic fermentation characteristics of different types of protein-rich substrates in food waste to produce volatile fatty acids. *Bioresour. Technol.* **2017**, *227*, 125–132. [CrossRef]
78. Gollakota, A.; Savage, P.E. Hydrothermal liquefaction of model food waste biomolecules and ternary mixtures under isothermal and fast conditions. *ACS Sustain. Chem. Eng.* **2018**, *6*, 9018–9027. [CrossRef]
79. Li, Y.; Liu, H.; Xiao, K.; Jin, M.; Xiao, H.; Yao, H. Combustion and pyrolysis characteristics of hydrochar prepared by hydrothermal carbonization of typical food waste: Influence of carbohydrates, proteins, and lipids. *Energy Fuels* **2019**, *34*, 430–439. [CrossRef]
80. Kostyukevich, Y.; Vlaskin, M.; Borisova, L.; Zherebker, A.; Perminova, I.; Kononikhin, A.; Popov, I.; Nikolaev, E. Investigation of bio-oil produced by hydrothermal liquefaction of food waste using ultrahigh resolution Fourier transform ion cyclotron resonance mass spectrometry. *Eur. J. Mass Spectrom.* **2018**, *24*, 116–123. [CrossRef]
81. Posmanik, R.; Cantero, D.A.; Malkani, A.; Sills, D.L.; Tester, J.W. Biomass conversion to bio-oil using sub-critical water: Study of model compounds for food processing waste. *J. Supercrit. Fluids* **2017**, *119*, 26–35. [CrossRef]
82. Tasca, A.L.; Puccini, M.; Gori, R.; Corsi, I.; Galletti, A.M.R.; Vitolo, S. Hydrothermal carbonization of sewage sludge: A critical analysis of process severity, hydrochar properties and environmental implications. *Waste Manag.* **2019**, *93*, 1–13. [CrossRef] [PubMed]
83. Sharma, H.B.; Dubey, B.K. Co-hydrothermal carbonization of food waste with yard waste for solid biofuel production: Hydrochar characterization and its pelletization. *Waste Manag.* **2020**, *118*, 521–533. [CrossRef] [PubMed]
84. Wang, Q.; Wu, S.; Cui, D.; Pan, S.; Xu, F.; Xu, F.; Wang, Z.; Li, G. Co-hydrothermal carbonization of corn stover and food waste: Characterization of hydrochar, synergistic effects, and combustion characteristic analysis. *J. Environ. Chem. Eng.* **2022**, *10*, 108716. [CrossRef]
85. Wang, L.; Chi, Y.; Du, K.; Zhou, Z.; Wang, F.; Huang, Q. Hydrothermal treatment of food waste for bio-fertilizer production: Formation and regulation of humus substances in hydrochar. *Sci. Total Environ.* **2022**, *838*, 155900. [CrossRef]
86. Xu, Y.; Wang, B.; Ding, S.; Zhao, M.; Ji, Y.; Xie, W.; Feng, Z.; Feng, Y. Hydrothermal carbonization of kitchen waste: An analysis of solid and aqueous products and the application of hydrochar to paddy soil. *Sci. Total Environ.* **2022**, *850*, 157953. [CrossRef]
87. Cavali, M.; Libardi, N., Jr.; de Sena, J.D.; Woiciechowski, A.L.; Soccol, C.R.; Filho, P.B.; Bayard, R.; Benbelkacem, H.; de Castilhos, A.B., Jr. A review on hydrothermal carbonization of potential biomass wastes, characterization and environmental applications of hydrochar, and biorefinery perspectives of the process. *Sci. Total Environ.* **2023**, *857*, 159627. [CrossRef]
88. Hwang, I.-H.; Aoyama, H.; Matsuto, T.; Nakagishi, T.; Matsuo, T. Recovery of solid fuel from municipal solid waste by hydrothermal treatment using subcritical water. *Waste Manag.* **2012**, *32*, 410–416. [CrossRef]
89. Berge, N.D.; Ro, K.S.; Mao, J.; Flora, J.R.V.; Chappell, M.A.; Bae, S. Hydrothermal Carbonization of Municipal Waste Streams. *Environ. Sci. Technol.* **2011**, *45*, 5696–5703. [CrossRef]
90. Lu, X.; Berge, N.D. Influence of feedstock chemical composition on product formation and characteristics derived from the hydrothermal carbonization of mixed feedstocks. *Bioresour. Technol.* **2014**, *166*, 120–131. [CrossRef]
91. Pala, M.; Kantarli, I.C.; Buyukisik, H.B.; Yanik, J. Hydrothermal carbonization and torrefaction of grape pomace: A comparative evaluation. *Bioresour. Technol.* **2014**, *161*, 255–262. [CrossRef] [PubMed]
92. Lu, J.; Liu, Z.; Zhang, Y.; Savage, P.E. Synergistic and antagonistic interactions during hydrothermal liquefaction of soybean oil, soy protein, cellulose, xylose, and lignin. *ACS Sustain. Chem. Eng.* **2018**, *6*, 14501–14509. [CrossRef]
93. Fox, J.T.; Zook, A.N.; Freiss, J.; Appel, B.; Appel, J.; Ozsuer, C.; Sarac, M. Thermal conversion of blended food production waste and municipal sewage sludge to recoverable products. *J. Clean. Prod.* **2019**, *220*, 57–64. [CrossRef]
94. Wang, T.; Zhai, Y.; Li, H.; Zhu, Y.; Li, S.; Peng, C.; Wang, B.; Wang, Z.; Xi, Y.; Wang, S.; et al. Co-hydrothermal carbonization of food waste-woody biomass blend towards biofuel pellets production. *Bioresour. Technol.* **2018**, *267*, 371–377. [CrossRef] [PubMed]
95. Zheng, C.; Ma, X.; Yao, Z.; Chen, X. The properties and combustion behaviors of hydrochars derived from co-hydrothermal carbonization of sewage sludge and food waste. *Bioresour. Technol.* **2019**, *285*, 121347. [CrossRef]
96. Mazumder, S.; Saha, P.; Reza, M.T. Co-hydrothermal carbonization of coal waste and food waste: Fuel characteristics. *Biomass Convers. Biorefinery* **2020**, *12*, 3–13. [CrossRef]
97. Su, H.; Liao, W.; Wang, J.; Hantoko, D.; Zhou, Z.; Feng, H.; Jiang, J.; Yan, M. Assessment of supercritical water gasification of food waste under the background of waste sorting: Influences of plastic waste contents. *Int. J. Hydrog. Energy* **2020**, *45*, 21138–21147. [CrossRef]
98. Chen, W.T.; Zhang, Y.H.; Zhang, J.X.; Yu, G.; Schideman, L.C.; Zhang, P.; Minarick, M. Hydrothermal liquefaction of mixed-culture algal biomass from wastewater treatment system into bio-crude oil. *Bioresour. Technol.* **2014**, *152*, 130–139. [CrossRef]

99. Santos Santana, M.; Pereira Alves, R.; da Silva Borges, W.M.; Francisquini, E.; Guerreiro, M.C. Hydrochar production from defective coffee beans by hydrothermal carbonization. *Bioresour. Technol.* **2020**, *300*, 122653. [CrossRef]
100. McGaughy, K.; Toufiq Reza, M. Hydrothermal carbonization of food waste: Simplified process simulation model based on experimental results. *Biomass Convers. Biorefin.* **2017**, *8*, 283–292. [CrossRef]
101. Venna, S.; Sharma, H.B.; Reddy, P.H.P.; Chowdhury, S.; Dubey, B.K. Landfill leachate as an alternative moisture source for hydrothermal carbonization of municipal solid wastes to solid biofuels. *Bioresour. Technol.* **2021**, *320 Pt B*, 124410. [CrossRef] [PubMed]
102. Kim, H.W.; Cho, W.; Lee, J.-Y. Characterization of bio-coal made via hydrothermal carbonization of mixed organic waste. *J. Korea Soc. Waste Manag.* **2020**, *37*, 93–101. [CrossRef]
103. Liu, H.; Ding, J.; Chen, L.; Ding, L. A novel fluorescence assay based on self-doping biomass carbon dots for rapid detection of dimethoate. *J. Photochem. Photobiol. A* **2020**, *400*, 112724. [CrossRef]
104. Zhou, Y.; Liu, Y.; Li, Y.; He, Z.; Xu, Q.; Chen, Y.; Street, J.; Guo, H.; Nelles, M. Multicolor carbon nanodots from food waste and their heavy metal ion detection application. *RSC Adv.* **2018**, *8*, 23657–23662. [CrossRef] [PubMed]
105. Wu, G.; Feng, M.; Zhan, H. Generation of nitrogen-doped photoluminescent carbonaceous nanodots via the hydrothermal treatment of fish scales for the detection of hypochlorite. *RSC Adv.* **2015**, *5*, 44636–44641. [CrossRef]
106. Yao, G.; Guo, Y.; Le, Y.; Jin, B.; He, R.; Zhong, H.; Jin, F. Energy valorization of food waste: Rapid conversion of typical polysaccharide components to formate. *Ind. Eng. Chem. Res.* **2020**, *59*, 17069–17075. [CrossRef]
107. Chengli, Y.; Jinmiao, Z. Mineralization of hydroxyapatite induced by eggshell as calcium source with hydrothermal synthesis method. *Crystallogr. Rep.* **2020**, *65*, 1242–1247. [CrossRef]
108. Alif, M.F.; Aprillia, W.; Arief, S. A hydrothermal synthesis of natural hydroxyapatite obtained from Corbicula moltkiana freshwater clams shell biowaste. *Mater. Lett.* **2018**, *230*, 40–43. [CrossRef]
109. Panahi-Kalamuei, M.; Amiri, O.; Salavati-Niasari, M. Green hydrothermal synthesis of high quality single and few layers graphene sheets by bread waste as precursor. *J. Mater. Res. Technol.* **2020**, *9*, 2679–2690. [CrossRef]
110. Zhang, F.; Wu, Q. Functional materials development from kitchen waste. *Procedia Environ. Sci.* **2012**, *16*, 70–74. [CrossRef]
111. Chung, W.; Oh, M.; Cho, W.; Park, S.-K.; Lee, J.-Y. A study on the adsorption of heavy metal with food waste bio-char using hydrothermal carbonization (HTC). *J. Korea Soc. Waste Manag.* **2016**, *33*, 137–144. [CrossRef]
112. Feng, Y.; Sun, H.; Han, L.; Xue, L.; Chen, Y.; Yang, L.; Xing, B. Fabrication of hydrochar based on food waste (FWHTC) and its application in aqueous solution rare earth ions adsorptive removal: Process, mechanisms and disposal methodology. *J. Clean. Prod.* **2019**, *212*, 1423–1433. [CrossRef]
113. Jin, F.; Zhang, G.; Jin, Y.; Watanabe, Y.; Kishita, A.; Enomoto, H. A new process for producing calcium acetate from vegetable wastes for use as an environmentally friendly deicer. *Bioresour. Technol.* **2010**, *101*, 7299–7306. [CrossRef] [PubMed]
114. Ren, K.; Zhang, W.; Cao, S.; Wang, G.; Zhou, Z. Carbon-based Fe_3O_4 nanocomposites derived from waste pomelo peels for magnetic solid-phase extraction of 11 triazole fungicides in fruit samples. *Nanomaterials* **2018**, *8*, 302. [CrossRef]
115. Fan, J.; De Bruyn, M.; Zhu, Z.; Budarin, V.; Gronnow, M.; Gomez, L.D.; Macquarrie, D.; Clark, J. Microwave-enhanced formation of glucose from cellulosic waste. *Chem. Eng. Process.* **2013**, *71*, 37–42. [CrossRef]
116. Kannan, S.; Gariepy, Y.; Raghavan, V. Optimization of enzyme hydrolysis of seafood waste for microwave hydrothermal carbonization. *Energy Fuels* **2015**, *29*, 8006–8016. [CrossRef]
117. Kaur, G.J.; Kumar, D.; Orsat, V.; Singh, A. Assessment of carrot rejects and wastes for food product development and as a biofuel. *Biomass Convers. Biorefinery* **2020**, *12*, 757–768. [CrossRef]
118. Yu, X.; Yin, J.; Wang, K.; Shen, D.; Long, Y.; Chen, T. Enhancing food waste hydrolysis and the production rate of volatile fatty acids by prefermentation and hydrothermal pretreatments. *Energy Fuels* **2016**, *30*, 4002–4008. [CrossRef]
119. Chen, W.-T.; Zhang, Y.; Lee, T.H.; Wu, Z.; Si, B.; Lee, C.-F.F.; Lin, A.; Sharma, B.K. Renewable diesel blendstocks produced by hydrothermal liquefaction of wet biowaste. *Nat. Sustain.* **2018**, *1*, 702–710. [CrossRef]
120. Kaushik, R.; Parshetti, G.K.; Liu, Z.; Balasubramanian, R. Enzyme-assisted hydrothermal treatment of food waste for co-production of hydrochar and bio-oil. *Bioresour. Technol.* **2014**, *168*, 267–274. [CrossRef]
121. Shao, Y.; Long, Y.; Zhou, Y.; Jin, Z.; Zhou, D.; Shen, D. 5-Hydroxymethylfurfural production from watermelon peel by microwave hydrothermal liquefaction. *Energy* **2019**, *174*, 198–205. [CrossRef]
122. Mahssin, Z.Y.; Hassan, N.A.; Yaacob, H.; Puteh, M.H.; Amin, N.A.S.; Zainol, M.M.; Hainin, M.R. Characterization of asphalt binder containing hydrothermal liquefied composition extracted from food waste. *IOP Conf. Ser. Earth Environ. Sci.* **2019**, *220*, 012013. [CrossRef]
123. Chua, G.K.; Tan, F.H.Y.; Chew, F.N.; Mohd-Hairul, A.R. Nutrients content of food wastes from different sources and its pre-treatment. *AIP Conf. Proc.* **2019**, *2124*, 020031. [CrossRef]
124. Qiao, W.; Yan, X.; Ye, J.; Sun, Y.; Wang, W.; Zhang, Z. Evaluation of biogas production from different biomass wastes with/without hydrothermal pretreatment. *Renew. Energy* **2011**, *36*, 3313–3318. [CrossRef]
125. Jia, X.; Xi, B.; Li, M.; Xia, T.; Hao, Y.; Liu, D.; Hou, J. Evaluation of biogasification and energy consumption from food waste using short-term hydrothermal pretreatment coupled with different anaerobic digestion processes. *J. Clean. Prod.* **2017**, *152*, 364–368. [CrossRef]
126. Redwood, M.D.; Orozco, R.L.; Majewski, A.J.; Macaskie, L.E. An integrated biohydrogen refinery: Synergy of photofermentation, extractive fermentation and hydrothermal hydrolysis of food wastes. *Bioresour. Technol.* **2012**, *119*, 384–392. [CrossRef]

127. Liu, J.; Wang, X.; Nie, X.; Li, R.; Song, M. In-situ emission characteristics of odorous gases from two food waste processing plants. *J. Mater. Cycles Waste Manag.* **2013**, *15*, 510–515. [CrossRef]
128. Villa Montoya, A.C.; da Silva Mazareli, R.C.; Silva, E.L.; Varesche, M.B.A. Improving the hydrogen production from coffee waste through hydrothermal pretreatment, co-digestion and microbial consortium bioaugmentation. *Biomass Bioenergy* **2020**, *137*, 105551. [CrossRef]

processes

MDPI

Review

Oxygenated and Nitrated Polycyclic Aromatic Hydrocarbons: Sources, Quantification, Incidence, Toxicity, and Fate in Soil—A Review Study

Wei Cao [1,2], Jing Yuan [3], Shuying Geng [2], Jing Zou [2], Junfeng Dou [2,*] and Fuqiang Fan [4,*]

1 School of Environmental and Municipal Engineering, Tianjin Chengjian University, Tianjin 300384, China
2 College of Water Sciences, Beijing Normal University, Beijing 100875, China
3 National Engineering Laboratory for Lake Pollution Control and Ecological Restoration, State Key Laboratory of Environmental Criteria and Risk Assessment, Chinese Research Academy of Environmental Sciences, Beijing 100012, China
4 Advanced Institute of Natural Sciences, Beijing Normal University at Zhuhai, Zhuhai 519087, China
* Correspondence: doujf@bnu.edu.cn (J.D.); ffan02@bnu.edu.cn (F.F.)

Abstract: The genotoxicity, mutagenesis, and carcinogenic effects of polycyclic aromatic hydrocarbon (PAH) derivatives may exceed the parent PAHs. However, their influence on the soil environment has not been explored to a large extent. Oxygenated polycyclic aromatic hydrocarbons (OPAHs) and nitrated polycyclic aromatic hydrocarbons (NPAHs) are typical polar substituted compounds. We offer a review of the literature on the sources, quantification, incidence, toxicity, and transport of these compounds in soil. Although their environmental concentrations are lower than those of their parent compounds, they exert higher toxicity. Both types of substances are basically related to carcinogenesis. OPAHs are not enzymatically activated and can generate reactive oxygen species in biological cells, while NPAHs have been shown to be mutagenic, genotoxic, and cytotoxic. These compounds are largely derived from the transformation of PAHs, but they behave differently in soil because of their higher molecular weight and dissimilar adsorption mechanisms. Therefore, specialized knowledge of model derivatives is required. We also made recommendations for future directions based on existing research. It is expected that the review will trigger scientific discussions and provide a research basis for further study on PAH derivatives in the soil environment.

Keywords: soil; PAH; OPAH; NPAH; toxicity; fate and behavior

1. Introduction

Polycyclic aromatic hydrocarbons (PAHs) are ubiquitous organic pollutants. Since Blumer [1] first reported PAHs in soil, extensive studies have been conducted to assess PAH toxicity, environmental persistence, and stability. More than 30,000 PAHs have been described, and 16 were listed as priority pollutants by the United States Environmental Protection Agency. PAHs enter the environmental matrix mainly through atmospheric emissions [2], and the global emissions of 16 priority PAHs are estimated to be 499 Gg [3]. PAHs are transported by air currents and have been recorded even in remote areas such as Greenland and Antarctica [4,5].

These persistent organic pollutants have been well studied in the atmosphere, water, and soil and can form related organic pollutants, termed substituted polycyclic aromatic hydrocarbons (SPAHs). PAHs and atmospheric oxides (e.g., O_3, NO_2, OH) can easily form SPAHs, which then migrate to soil and can transform into nitrated PAHs (NPAHs), oxygenated PAHs (OPAHs), and chlorinated PAHs. Polar derivatives of OPAHs and NPAHs have received recent attention [6,7]; in these compounds, the hydrogen on the aromatic ring is replaced by carbonyl and nitro functional groups [7–9]. The formation and release of these polycyclic aromatic compounds (PACs) are mainly caused by the incomplete combustion of fossil fuels, biofuels, and biomass. Multiple OPAHs and NPAHs have been reported to exert greater toxicity and to be more stable than their parent PAHs [10], although their sources are similar or identical. Therefore, these two types of organic pollutants are often used in comparative studies [11,12].

OPAHs and NPAHs are typical polar derivatives of PAHs and have been studied in various environmental matrices. Apart from the incomplete combustion of fossil fuels and biomass, these compounds can be formed by photochemical reactions in the air [13]. Most research on NPAHs and OPAHs has been conducted in their particulate airborne phase [14–17] and is relatively mature. Since water is a fluid and unstable medium, the incidence of waterborne SPAHs is relatively low. Soil is both a source and a sink of atmospheric pollutants and especially for PACs produced by human activities, through both dry/wet deposition and direct air–soil exchange. A study conducted in the UK estimated that as many as 90% of PACs were stored in soil [18]. In recent years, more and more attention has been paid to soilborne PAH derivatives, both the less-studied polar derivatives and pure OPAHs and NPAHs, and their toxicity [19,20]. Because polar functional groups have the effect of soil minerals and are affected by soil chemistry, the adsorption mechanisms of OPAHs and NPAHs may be more diverse than those of PAHs [21,22]. In short-term incubation experiments of fertile soil, the microbial transformation of parent PAHs was shown to produce OPAHs [23]. The water solubility of OPAHs and NPAHs is higher than the solubility of related PAHs, which generally results in faster vertical leaching in soil [24,25] but differing soil dynamics. In this review, we compare and summarize the research into OPAHs and NPAHs in the soil to provide a systematic theoretical framework for these new organic pollutants (Figure 1).

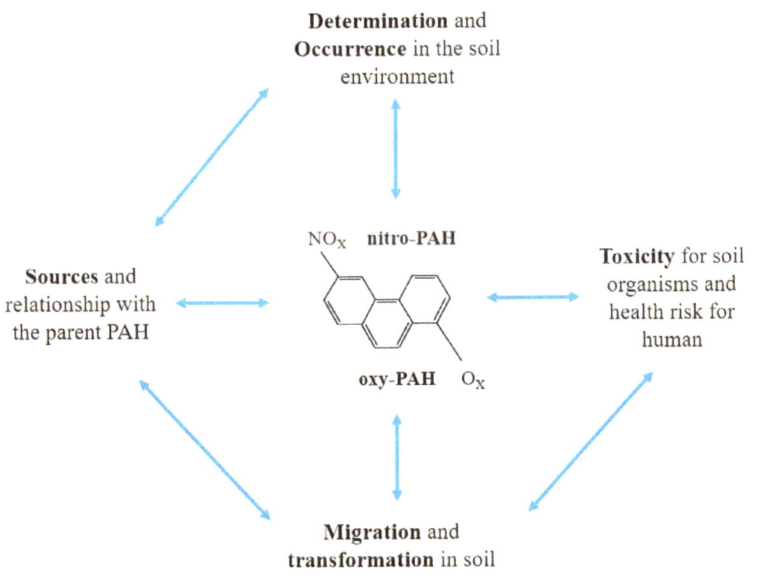

Figure 1. Structure diagram of this review.

This paper is a comprehensive review of journals, books, and related government reports about OPAH and NPAH in the soil environment. It summarizes the sources, detection, occurrence, migration and transformation, toxicity, and health risks. According to the relevant content, representativeness, and the latest methodological progress, the recent studies on OPAHs and NPAHs in soil are included in this review. This can provide a research basis for studying the environmental behavior and fate of polar SPAHs.

2. Sources

Unlike PAHs, which are mainly produced from incomplete combustion sources of coal, biomass, and motor vehicle emissions, OPAHs and NPAHs can be released into the environment simultaneously with PAHs or be formed by the reaction of the parent PAH with free radicals such as OH and NO_3 or through homogeneous or heterogeneous reactions [26,27]. Both NPAHs and OPAHs are mainly formed as direct or indirect products of incomplete combustion, such as of diesel and gasoline engines, or from interactions of gaseous air pollutants. The SPAHs produced by human activities can enter the soil through dry/wet deposition and direct air–soil exchange or as litter from contaminated vegetation [28,29]. The concentrations of PAHs and OPAHs in soil have been shown to be closely related [9,25]; conversion of PAHs to OPAHs in soil involves biological activity driven by climate and soil characteristics such as pH, nutrients, and levels of organic matter [30]. Higher biological activity is expected to result in a higher soil OPAH/PAH ratio [23]. The higher water solubility of OPAHs and NPAHs typically leads to faster vertical leaching in soil [22,24,25], and although the polar functional groups of OPAHs hinder soil transport, they can also bind more easily to soil solids [31].

Soil has great storage capacity for PAHs and their derivatives since soil organic matter is the main adsorbent of PACs [32]. Gas/particle distribution, adsorption-desorption, volatilization, abiotic degradation, biodegradation, leaching, and bioconcentration are important dissipation processes that affect the concentration and spatial distribution of PACs in soil [25,33]. These processes are affected by the physical and chemical properties of PACs, such as the octanol–water partition coefficient, octanol-air partition coefficient, and molecular weight [34]. Soil characteristics such as carbon concentration, moisture content, texture, structure, mineralogy, porosity, and enzyme and microbial properties are also important factors in the fate of OPAHs and NPAHs [23,35]. Climate factors such as temperature, precipitation, monsoon cycle, and hours of sun exposure can also affect the formation, concentrations, and composition of OPAHs and NPAHs in soil. Overall, the soil concentrations and spatial distribution of these compounds cannot be predicted from the degree of human activity.

Measurements of NPAH and OPAH concentrations in the soil must take into account isolation, leaching, volatilization, degradation, and bioconcentration [33]. Once deposited into the soil, these compounds will mainly adsorb to organic matter and remain in the topsoil or be transferred elsewhere through leaching, bioturbation, or colloid-assisted transportation [36]. However, because polar functional groups are affected by soil chemistry, the adsorption mechanisms of OPAHs and NPAHs may be more diverse [22]. Additionally, soil biodegradation of deposited PAHs, photodegradation, thermal degradation, and reactions catalyzed by soil inorganic constituents may all lead to the accumulation of OPAHs and NPAHs [23,37,38]. Microorganism action is another key factor in OPAH and NPAH accumulation. For OPAHs, the situation may be even more special. Because through the bioremediation process, the conversion of parent PAHs to OPAHs may occur, and PAHs containing mainly hydroxyl and carbonyl groups are generated [39]. The concentration of PAHs in the soil can be effectively reduced through the absorption and utilization of microorganisms [9,40], but it is not completely metabolized. The final product may be more toxic, in this regard, has not received sufficient attention. The formation of OPAHs by soil microorganisms has been studied, but the mechanism is still unclear.

3. Quantification

Polar derivatives of PAHs may exert greater toxicity than their parent aromatic compounds and may be enriched in contaminated soil but are rarely included in routine monitoring and risk assessment procedures because of a lack of regulations and standardized analysis methods. However, the more polar SPAHs have garnered research attention over the past few years. Studies in soil have mainly focused on the simultaneous quantification of PAHs and their derivatives [16,41]. Wilcke et al. [42] measured the sum totals of 29 PAHs, 15 OPAHs, and four azaarenes simultaneously in surface soil in Argentina, with an average recovery rate of 92%. Wei et al. [43] measured the sum totals of 29 PAHs, 15 SPAHs, 4 OPAHs, and 11 NPAHs in surface soil and road dust in China using the methods of Bandowe et al. [44]. Bandowe and colleagues measured PAHs, OPAHs, and NPAHs in urban street dust in Africa, with recovery rates in the range of 53%–152% [45].

There is no standardized method for analyzing polar PAHs in soil, and therefore the methods vary between study groups. Typically, the analysis is based on improved PAHs pre-processing and testing methods or based on airborne particle analysis methods. It is mainly composed of three steps (Figure 2): The first step is to extract the collected soil samples and then to carry out purification and fractionation steps. The third step is to identify and quantify the target compound. Sampling the soil from the surface or deeper depths, freezing or air-drying (freeze-drying) the soil sample before storing it in the refrigerator. If the soil is stored without pre-processing, a desiccant (Na_2SO_4, diatomaceous earth) can be added to remove the water before or after extracting the chemicals. Soxhlet extraction, accelerated solvent extraction, or ultrasonic extraction are used to remove organic matter from the soil, and solvents differ by the type required for the procedure. The complexity of the soil substrate typically requires additional purification [25,41]; this is because the concentration of SPAHs in the environment is generally lower than that of their parent PAHs and the methods are less sensitive to detecting these compounds. To improve the chromatographic signal quality and signal-to-noise ratio, the interference must be minimized. Solid-phase extraction columns or self-packed purification columns should be used to fractionate the extract to remove impurities and separate the target contaminants. The ratio and order of solvents used for elution will also differ. Internal and external standards are used to quantify targets with improved accuracy [46–48]. Additionally, prior to column fractionation, liquid–liquid extraction can be used to separate SPAHs with basic properties from the extract to improve the analysis of these compounds further.

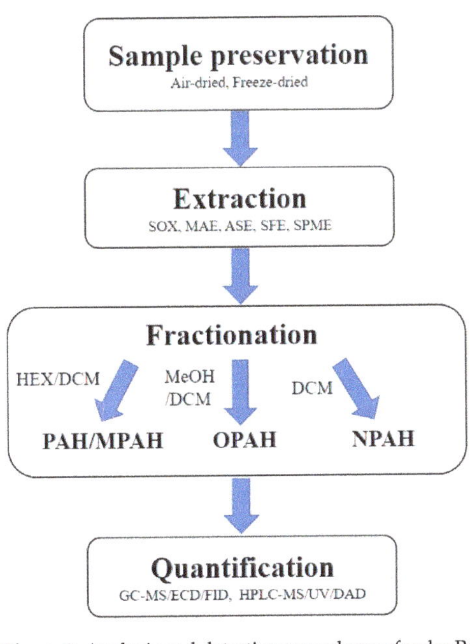

Figure 2. Analysis and detection procedures of polar PAHs.

Finally, the fractions after rotary evaporation and drying by nitrogen, the fractions are typically analyzed by gas or liquid chromatography, often with mass spectrometry. Target quantification is based on the ratio of the analytical peak area to the peak area of the internal standard [49]. Liquid chromatography sometimes harnesses diode arrays or fluorescence;

liquid chromatography with mass spectrometry uses atmospheric pressure chemical ionization or atmospheric pressure photoionization, while gas chromatography-mass spectrometry uses single or triple quadrupole instruments with electron ionization. To analyze OPAHs by gas chromatography, a derivatization step is generally included to increase the volatility and thermal stability of the molecule and reduce its surface interaction. To analyze NPAHs, mass spectrometry is typically operated in negative ion mode to increase the sensitivity of the target. For quality assurance, we recommend using the same analytical method for SPAHs and measuring blank or spiked soil samples in tandem.

4. Incidence

Environmental PAHs typically accumulate in the soil, which may also be the main storage for SPAHs. Research has mainly focused on industrial sites such as power plants, coking plants, and wood mills [9,20,50]. There are also some reports on SPAHs in urban soils. Work on OPAHs and NPAHs in soil has measured surface soil, road dust, and estimated human health risks using computational models. Because PAH derivatives are abundant, SPAHs have garnered research attention. Brorström-Lundén et al. [51] first reported that the background values of OPAHs and NPAHs in soil differed from concentrations in urban areas. After that, more scholars studied OPAHs and NPAHs and found differences in their soil concentrations between land use types (Table 1). Wilcke et al. [42] reported the concentration distribution and correlation between parent compounds and SPAHs in surface soils in Argentina, finding close associations between parent PAHs and their derivatives. Bandowe et al. [44] compared the soil concentrations of 15 OPAHs and 4 NPAHs in Thailand and found relationships with trace metals and that the concentrations and compositions of OPAHs and their parent compounds were similar, but that the ratio of PAHs was higher, perhaps owing to strong local microbial activity and photoactivity. Wei et al. [43] studied road dust and topsoil in Xi'an, China, and reported that PAHs and OPAHs but not NPAHs originated mainly in combustion activities; the enrichment factor in dust and soil was correlated with log octanol–air partition coefficient and log octanol-water partition coefficient. A study in Ghana [45] reported the transformation of PAHs into SPAH derivatives and the higher toxicity of OPAHs and NPAHs. Sun et al. [52] measured the concentrations of 11 NPAHs and 4 OPAHs in agricultural soils of 26 provinces in eastern China and reported that NPAH levels were higher than those of OPAHs, but we believe that this study is not representative. The reason is that the number of the two types of substances studied is quite different. Yadav et al. [53] measured SPAH concentrations in city dust in Nepal and reported that NPAH levels were 2–5 times higher than the levels of parent PAHs. They also concluded that the main exposure route in humans was dermal contact. A study in surface soils in Newcastle, Australia [54] indicated that polar and parent PAHs were widespread. Although it is an industrial city, its overall carcinogenic risk is negligible. According to the research situation of scholars, we mainly compared the occurrence of 14 OPAHs and 11 NPAHs in soil. According to the research situation of scholars, we mainly compared the occurrence of 14 OPAHs and 11 NPAHs in the soil; see Tables 2 and 3 for details.

4.1. OPAHs

For OPAHs, this review also lists a comparison table of different types of single substances around the world (Table 2). Regardless of soil type, the abundance and concentrations of 9-fluorenone are relatively high, which may be closely related to the fate and behavior of its parent PAH. OPAH/PAH (1-INDA/FLUO, 9-FLO/FLUO, 1,2-ACQ/ACENY, and 1,2-ACQ/ACEN) in the topsoil and subsoil in an industrial zone in Uzbekistan ANQ/ANTH) > 1, the OPAHs concentration in the soil is higher than the parent PAHs. The significantly positive correlations between carbonyl-OPAHs and trace metals in industrial waste gas support a common source for PAHs, carbonyl-OPAHs, and trace metals [55]. Higher OPAH concentrations have been associated with human activities, such as low-temperature combustion [42], and the lighter OPAH is detected in the south, while more OPAHs were detected in the north where human activities are frequent, from north to south 9,10-ANQ/Ant, indicating that it may originate from low-temperature combustion. In addition, the average concentration of ∑OPAHs in street dust in Kumasi, Ghana, is 833 ng g^{-1}. Furthermore, the similar spatial trend of OPAH and alkyl + parent PAHs concentrations is determined by the strong and significantly positive difference between ∑OPAH and Σ alkyl + parent PAHs. The relevance is further confirmed [45]; the composition mode of the OPAH in the street dust of Kumasi mainly consists of 9,10-ANQ, B(A)FLUone, and 9-FLO. The largest amount of 9-FLO was detected in the soil of Newcastle, a famous industrial city in Australia, with a range of 42.8–1818.1 ng g^{-1} [54]. Among different land types in the city, the most abundant OPAH is 9-FLO, which, respectively, accounts for 54.5, 44.9, 37.7, and 20.0% of the 7 types of OPAHs in residential, smoking, recreational and industrial areas, followed by 1,4-NQ that accounts for 18.0%, 25.3%, 32.7%, and 46.4% of industrial, smoking, residential and recreational soils accordingly. In industrial soils, BANTone is the most common oxy-PAH, covering 36.6% of 7 oxy-PAHs. However, its contribution to the soil of smoking areas, residential areas, and entertainment venues is relatively small, with 4.1%, 2.2%, and 4.9%, respectively. In soil, OPAH mobility is only slightly higher than that of the low-molecular-weight parent PAHs, which is probably due to complex interactions between OPAHs and PAHs in soil.

4.2. NPAHs

From the perspective of 11 NPAHs that scholars have studied more in recent years, the concentration ranges of substitutes in different land use types do not remain the same (Table 3). Wei et al. [43] investigated the concentrations of 11 NPAHs in the surface soils of Xi'an suburbs. The average concentration was 118 ± 52 ng g^{-1}, and the higher concentrations were 2 + 3-NFLA and 2,7-DNFLU. Through conducting the analysis with the combustion tracer, it is found that these substances are not derived from combustion activities. The total concentration of 16 NPAHs in road dust in the four cities of Nepal is 2–5 times lower than that of the parent PAHs [56]. 2-NFLUA/1-NPYR represents the secondary contribution from photochemical reactions. Sun et al. [52] analyzed 11 types of NPAHs in the soil of agricultural land in 26 provinces located in Eastern China, and then it was shown that except for 5 types of NPAH, which have a low detection rate. All other substances are logarithmically distributed. Individual NPAH is mostly higher than the soil concentration in other cities. In the meanwhile, from the current study, it was found that higher Kp can make the deposition flux of NPAHs from the air. There are not many studies on polar PAHs in soil, mainly from the above scholars. It can be found that although more scholars have participated in the research of OPAHs and NPAHs, there are still big problems: First of all, there are no unified standards and regulations for the standard samples of the two types of pollutants, which leads to large differences in the research materials of scholars, and there is no way to refer; Secondly, the research on PAC in soil mostly focuses on the occurrence and the associated risk assessment, ignoring formation mechanisms and the dynamics of migration into deeper levels of subsoil. Standardizing the methods and developing techniques for studying both formation mechanisms and migration dynamics should be the next steps in the field.

Table 1. The occurrence of OPAHs and NPAHs in different types of soils.

Region	Land Use	∑OPAHs (ng g^{-1})	# of Oxygen-PAH	∑NPAHs (ng g^{-1})	# of Nitro-PAH	References
Gårdsjön, Sweden Göteborg, Sweden	background area urban-diffuse	108 42–466	8	21 11–286	9	[51]
Patagonian, Argentina	steppe	2.4–38	15	0.05–124	4	[42]
Bangkok, Thailand	urban	12–269	15	0.1–31	4	[44]
Xi'an, China	suburban	854 ± 447	15	118 ± 52	11	[43]
Kumasi, Ghana	urban	833 (57–4202)	10	73 (3–240)	4	[45]
East China	agriculture	9 ± 8	4	50 ± 45	11	[52]
Kathmandu, Nepal Pokhara, Nepal Birgunj, Nepal Biratnagar, Nepal	urban	99 (95–384) 95 (92–433) 115 (93–348) 606 (227–1150)	3	646 (439–3930) 568 (409–2030) 832 (451–1750) 2100 (586–24,100)	16	[53]
Newcastle, Australia	recreation industry smoking area resident	1515 3924 644 3649	7	211 602 239 391	3	[54]

Table 2. Research status of individual OPAH in soil.

Compound	Abbrev	Concentration	References
1-indanone	1-INDA	1.6–35 (industrial area), 0.05–0.54 (steppe), 32.4–59.7 (suburban), 10.5–56.1(street dust)	[42,43,45,55]
1,4-naphthoquinone	1,4-NQ	0.3–6.7 (industrial area), 2.2–10.7 (suburban), 157.7–1443.1 (industrial heritage)	[43,54,55]
1-naphthaldehyde	1-NLD	1.4–20.6 (industrial area), 0.16 (steppe), 7.9–35 (suburban), 2.6–37.5 (street dust)	[42,43,45,55]
2-biphenylcarboxaldehyde	2-BPCD	0.3–25 (industrial area), 2.15 (steppe), 10.9–22.8 (suburban)	[42,43,55]
9-fluorenone	9-FLO	14–810 (industrial area), 1.02–2.82 (steppe), 371–1243.8 (suburban), 33.2–214.4 (street dust), 0.3–10 (farmland), 42.8–1818.1 (industrial heritage)	[42,43,45,52,54,55]
1,2-acenaphthylenedione	1,2-ACQ	2.6–14 (industrial area), 1.2 (steppe), 2.4–6.7 (suburban)	[42,43,55]
9,10-anthraquinone	9,10-ANQ	12–959 (industrial area), 0.88–8.9 (steppe), 36.7–92.9 (suburban), 18.1–1241.7 (street dust), 0.9–26(farmland), 33.2–202.7 (industrial heritage)	[42,43,45,52,54,55]
4H-cyclopenta[d,e,f]phenanthrene-4-one	CPHENone	12.2–48 (suburban), 5.6–131.2 (street dust)	[43,45]
2-methyl-9,10-anthraquinone	2-MANQ	2.3–18 (steppe), 7.9–27.7 (suburban), 7.1–1602.9 (street dust)	[42,43,45]
benzo[a]fluorenone	B(A)FLUone	0.13–5.2 (steppe), 7.1–37 (suburban), 5.7–919.6 (street dust)	[42,43,45]
7H-benz[d,e]anthracene-7-one	BANTone	0.43–16 (steppe), 3.5–19.2 (suburban), 12.7–200.2 (street dust), 25.8–388.3 (industrial heritage)	[42,43,45,54]
benzo[a]anthracene-7,12-dione	7,12-B(A)A	20 (steppe), 12.1–58.1 (suburban), 0.2–8.4(farmland)	[42,43,52]
5,12-naphthacenequinone	5,12-NACQ	1.5–2.4 (steppe), 2.9–28.6 (suburban)	[42,43]
6H-benzo[c,d]pyrene-6-one	6-BPYRone	1.9–37 (steppe), 2–10.7 (suburban), 6.0–66.0 (street dust)	[42,43,45]

Table 3. Research status of individual NPAH in soil.

Compound	Abbrev	Concentration (ng g^{-1})	References
1-Nitronaphthalene	1-NNAPH	0.4–1.8 (suburban), 0.2–267 (farmland), 51.9–77.6 (urban dust), 49.6–205 (industrial heritage)	[43,52–54]
2-Nitrobiphenyl	2-NBP	0.4 (suburban), 0.3–66 (farmland)	[43,52]
5-Nitroacenaphthene	5-NACEN	0.5–3.5 (suburban), 0.7–7.7 (farmland), 35.3–47.8 (urban dust)	[43,52,53]
2-Nitrofluorene	2-NFLU	0.2–0.8 (suburban), 7.5–9.2 (farmland), 21.8–631 (urban dust), 88.7–196 (industrial heritage)	[43,52–54]
9-Nitrophenanthrene	9-NPHEN	0.4–1.9 (suburban), 0.4–17 (farmland), 7.7–84.1 (urban dust)	[43,52,53]
9-nitroanthracene	9-NANT	0.1–0.7 (suburban), 0.2–8.5 (farmland), 0.14–1.24 (urban dust), 21–271(industrial heritage)	[43,52–54]
(2 + 3)-Nitrofluoranthenes	2 + 3-NFLA	5.2–16.3 (suburban), 0.4–29 (farmland)	[43,52]
1-Nitropyrene	1-NPYR	2.3–6.0 (suburban), 8.8 (farmland)	[43,52]
2,7-Dinitrofluorene	2,7-DNFLU	13.0–39.0 (suburban), 2.0–41.0 (farmland), 28.1–44.9 (urban dust)	[43,52,53]
6-Nitrochrysene	6-NCHR	5.7–11.4 (suburban), 26.3–67.2 (urban dust)	[43,53]
6-Nitrobenzo[a]pyrene	6-NBaP	49.9–86.1 (suburban), 50.3–65.9 (urban dust)	[43,53]

5. Toxicity and Human Health Risk

Polar PACs are stable in the environment and can bioaccumulate, and exert toxic and ecotoxicological effects [9,57]. OPAHs and NPAHs are directly mutagenic, and some are estrogenic compounds and generate reactive oxygen species (ROS) responsible for oxidative stress (OS) [58,59]. OPAHs, NPAHs, or the environmental matrix extracts containing OPAHs and NPAHs, have similar or even more obvious toxicity than the related parent PAHs or the PAHs extracts [9,10,58]. The toxicity profile of OPAHs and NPAHs can be seen in Figure 3 outlines the toxic effects of OPAHs and NPAHs in soil.

Polar PAH has received more attention due to its toxicity, persistence, and wide mobility in the soil environment [60]. OPAH is an important class of polar PAHs. Compared with the parent PAH of the same ring, OPAHs have more complex carcinogenic effects due to the existence of negatively charged atoms [61]. Usually, these carcinogenic effects are related to non-enzymatic activation, which will promote the abnormal expression of genes and the mutagenesis of proteins. Similarly, certain NPAHs are also listed as human carcinogens by the International Agency for Research on Cancer (IARC) [62,63]. The substitution of nitrogen atoms in the fused ring structure impacts the physical, chemical, and toxicological properties of NPAHs. In soil, nitrogen may affect the absorption rate of NPAHs and their toxicity to soil organisms. Unlike homocyclic PAHs, the high reactivity of NPAHs and OPAHs may result in volatilization, low adsorption, and high degradation [64]. Relative to their parent PAHs, NPAHs, and OPAHs are weak bases, which can be protonated in soil solution, depending on their basicity and on soil pH [56,65]. In most natural soils, weak bases exist as molecules, and decreases in soil pH may produce the protonated form (cations). These cations can bind with soil cation exchange sites, thereby increasing their half-life and persistence [66].

5.1. Exposure

The exposure pathways are identical for polar PAHs and their parent compounds: water, air, soil, and food through ingestion, inhalation, skin, and eye contact [67–69]. Ingestion of PAHs has been associated more with known morbidities than with other routes of exposure in humans [70]. NPAHs and OPAHs in the air are produced during the combustion processes of heating, cooking, vehicular traffic, and manufacturing [71–73]. Inhalation is an important exposure route, and levels of NPAH biomarkers are higher in urban populations than in rural ones [74,75]. Inhalation of 1-nitropyrene produced squamous metaplasia and cytoplasmic changes in rats, with a no-observed-adverse-effect level of 0.5 mg/m^3. NPAHs have been detected in vegetables, fruits, tea, barbecued or baked meat, water, and crop soil [15,76]. Yaffe et al. [70] estimated that the acceptable daily intake of 1-nitropyrene in adults is approximately 9.7×10^{-7} mg/kg/day.

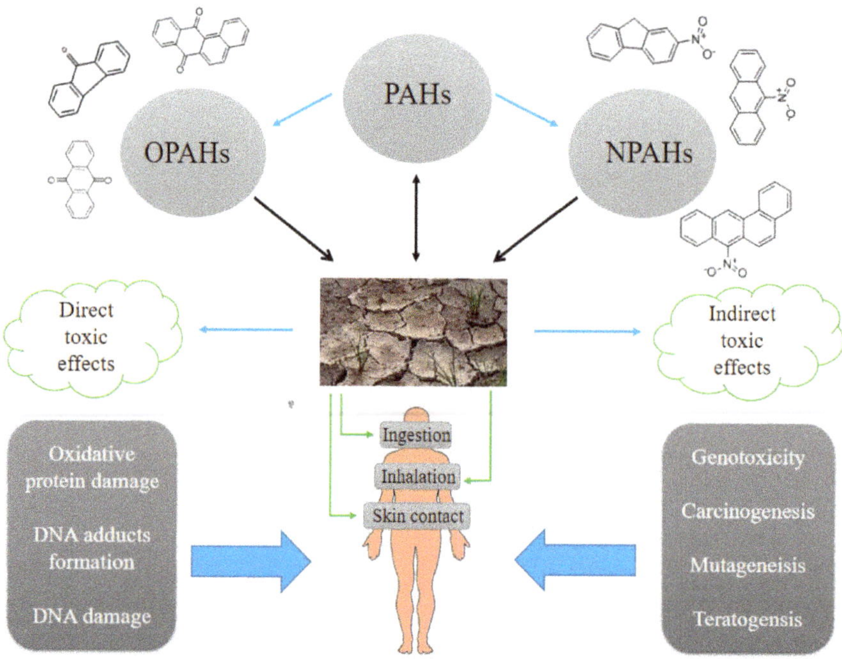

Figure 3. Toxic effects of OPAHs and NPAHs in soil.

5.2. Mutagenicity, Carcinogenicity, and Teratogenicity

There are many PAHs exposed through the soil, which pose a great threat to human health [77–79]. Although PAH does not cause direct DNA damage [80], it can often form highly reactive genetic metabolites through the "activation" effect of organisms in the soil. For example, quinone and anthraquinone, these intermediate products that have the potential to synthesize with DNA [81].

The toxicity of NPAHs in environmental media such as soil, sediments, and air has been studied extensively in multiple test systems. The mutagenicity of several NPAHs and OPAHs was tested in the Salmonella Ames assay [82,83]. For instance, it is shown by the salmonella mutagenicity test (Ames test) that the extracts of soils collected from several Japanese areas (n = 544) display mutagenicity. It is suspected that the mutagenicity of these soils is related to others and shown for NPAHs such as 3,6-dinitrobenzene, 1,3,6-trinitropyrene, 3,9-dinitroanthracene, and 3-nitrobenzanthrone [84–88]. These research results verify the mutagenicity, genotoxicity, carcinogenicity, and estrogen activity of the several NPAHs [73,89]. According to the published scientific research results, the human carcinogenic risk assessment workgroup of IARC has classified several SPAHs as the carcinogenic group. Some SPAHs have been assigned a toxic equivalency factor far higher than that of benzo[a]pyrene [90,91].

The mutagenicity of OPAHs and NPAHs in airborne particulates was confirmed in vitro [92] and can be double that of the parent PAHs [60]. In the research on human cell mutagenic agents in air-borne inhalable particles in northeast America, OPAH is reported as a highly mutagenic compound in vitro [92]. Although in vitro and in vivo effects of quinones have been described, little is known about the developmental toxicity of OPAHs. Naphthoquinone binds with biological macromolecules, and the alkylation and oxidation/reduction of benzoquinone will produce multiple toxic effects in vivo, including oxidative stress, acute cytotoxicity, immunotoxicity, and carcinogenicity [93,94]. Owing to the lack of toxic equivalency factors for NPAHs and OPAHs, it is impossible to correlate the contribution of exposure from the soil with human cancer risks. Instead, the carcinogenicity of polar PAHs is typically compared with that of the parent PAHs [52,56].

5.3. Stress and Genotoxicity

OPAH and NPAH have similar toxicity mechanisms, both of which are related to the ability to activate aromatic hydrocarbon receptors (AhR), and the level of AhR activation can be used as a simple toxicity indicator [83]. The ligand-binding group AhR is an intracellular ligand-activated transcription complex that can activate the transcription of genes related to cytochromes and glutathione S-converting enzymes. The specific action path of AhR is as follows: First, the activated AhR translocates to the nucleus and binds to the AhR nuclear transcription protein (Arnt) in the nucleus to form a heterodimer and then triggers the transcriptional modulation of genes containing heterogeneous response elements [62,95]. The special toxicity of OPAH is reflected in oxidative stress, which is mainly due to the imbalance between reactive oxygen species production and antioxidants [96]. OPAH toxicity may eventually lead to mitochondrial malformations and dysfunction, DNA breakdown, apoptosis, protein and lipid degradation, mutations, and cancer [97]. At the same time, it affects the regulation of gene expression and thus plays a role in tumors [98,99].

The in vivo mutagenicity of 2-nitro-fluorinated uranium was also shown in the soil nematode *Caenorhabditis elegans* [100]. The bioluminescent bacterium *Vibrio fischeri* has been used to investigate the short-term toxicity of NPAHs such as 5-nitro-acenephthene and 5-nitronaphthalene, showing that these compounds were more toxic than related PAHs [101]. In contrast, short-term exposure to 10 NPAHs in algae, Crustacea, or trees of the *Platycladus* genus did not yield evidence of toxicity. A statistically significant increase in genotoxicity was reported in a study that used the DT 40 chicken T lymphocyte bioassay to detect the genotoxicity of soil sample extracts after bioremediation [102]. Using a bioreactor to treat soil contaminated by PAH will increase the genotoxicity of the entire treatment cycle [47,102]. The increase in genotoxicity is attributed to the polar products produced by bioremediation. Due to the production of OPAH and other polar, toxic PAH conversion products, increased mutagenicity was observed after the aerobic bioremediation of soil contaminated by PAH [47]. Using air particle samples, the mutagenicity of OPAH and NPAH components from Beijing is twice that of the parent component [60]. Due to optimization, the bioavailability of soil pollutants may also be improved [103]. Because polar PAHs have high toxicity and potential genetic toxicity to microorganisms, more and more scholars have studied the developmental behavior, fate, and behavior in soil [50,104,105].

6. Transport and Transformation

Soil adsorption of polar PAHs affects their transport, fate, and ecotoxicity. Like other semi-volatile organic pollutants, OPAHs and NPAHs in soil are absorbed into organic matter, and black carbon through nonspecific hydrophobic interactions [28], and the adsorption strength can be described by the compounds' mobility. NPAHs have been shown to have lower mobility than that of their parent PAHs [70,83]. The clay fraction, in particular, is involved in the adsorption of polar PAHs, and the intensity of specific interaction with soil minerals and organic matter is thought to be equal to or higher than the nonspecific hydrophobic interaction [106]. The π–π interactions of the electron-deficient NPAHs and

OPAHs (electron acceptors) with the electron-rich soil organic matter and minerals (electron donors) account for the strong interaction. Other mechanisms include the complexation of a nitryl group and cation in clay (e.g., montmorillonite) or of siloxane-containing nitryl group and surface oxygen in clay (e.g., kaolinite) [106,107]. Polar PAHs exhibit a high adsorption efficiency to both topsoil and subsoil [19,106], and their concentrations are generally higher in subsoil if the topsoil has been treated with sewage sludge for agricultural purposes; in undisturbed soil. Polar PAHs have low water solubility and are extensively distributed into the solid soil phase, suggesting only negligible vertical through leaching [89]. OPAHs may be transported into the soil through inorganic and organic colloid-assisted transport [19,28,50]. The strong adsorption of NPAHs onto the solid soil phase might be influenced by wind and water erosion.

In addition, many scholars have focused on the biological migration and transformation of PACs in soil. Geng et al. [108] found that the fungi *Neocucurbitaria, Penicillium, Fusarium, Chrysosporium, Knufia, Filobasidium, Wallemia,* and *Rhodotorula* were identified as the keystone taxa, indicating that fungi also had important positions in the interdomain molecular ecological networks of different seasons. Acenaphthylene, benzo[b]fluoranthene, indeno [1,2,3-cd]perylene, benzo[g,h,i]pyrene, and 9-fluorenone were the key environmental factors driving the deterministic assembly processes of the interdomain microbial community in the contaminated soil. The combined bacterial, archaeal, and fungal community structure showed no significant seasonal changes under natural attenuation [11,109]. However, for single microbial communities, bacteria were more sensitive to seasonal variations than fungi and archaea. Similarly, bacteria have been reported to be sensitive to changes in environmental conditions in other studies [110–112]. The differences in sensitivity between bacteria, archaea, and fungi may be due to a large number of contaminant-degrading species in bacteria and the distinctive differences in cellular structure and physiological activity among them [108]. In addition, environmental filtering could govern microbial community assembly [113,114]. These findings extend our knowledge of interdomain microbial community assembly mechanisms and ecological patterns in natural attenuation and provide valuable guidance in associated biotransformation strategies.

As the generation process of PAH derivatives in the soil is relatively irreversible, with the conversion of more parent PAHs to SPAHs, the occurrence of SPAHs in the soil will be greatly accumulated. To our knowledge, no studies have been published on the gas-phase transport of OPAHs or NPAHs in soil, but this transport may be extremely limited because of the compounds' low volatility and strong adsorption to soil. The polar PAHs have low vapor pressure, high octanol-air partition coefficient, and strong adsorption onto soil; therefore, their volatility is estimated to be low. There are still few studies on the fate mechanism of OPAH and NPAH in soil, and further research is needed.

7. Summary and Outlook

This review focused on the state of knowledge in the detection, incidence, toxicity, and fate of OPAHs and NPAHs in soil. SPAHs represented by OPAHs and NPAHs are widespread in the soil environment, and their toxicity and mutagenic properties are more obvious, and their accumulation in organisms is also more complicated. Although their soil concentrations are generally lower than those of their parent compounds, OPAHs and NPAHs are persistent and may pose health risks. Unified testing standards and regulatory control of these compounds are urgently needed, as is research on their exchange between environmental media. The research on related pollutants in the soil also only focuses on the occurrence of surface soil or dust and health risks, and there is a lack of research on the exchange of other media with soil. The adsorption mechanism and vertical distribution in soil and groundwater, as well as the partition between soil solids and pore water, have not been reported. It is also necessary to conduct soil column experiments and infiltration instrument research to simulate its vertical migration in the soil. Due to the lack of relevant research data above, it is necessary to strengthen research to more clearly understand the fate and behavior of OPAHs and NPAHs in the soil environment. According to the reviewed articles, we can prove that the two types of substances exist and have higher mobility in the soil environment and may cause greater risks to organisms. It could be concluded that PACs should be included in monitoring plans and risk assessment models in the future.

Author Contributions: W.C.: Conceptualization, data curation, writing—original draft; J.Y.: investigation, methodology; S.G.: visualization; J.Z.: supervision, validation; J.D.: writing—review and editing; F.F.: writing—review and editing. All authors have read and agreed to the published version of the manuscript.

Funding: This study was supported by the National Key Research and Development Program of China (No. 2018YFC1800905).

Conflicts of Interest: The authors declare no conflict of interest.

References

1. Blumer, M. Benzpyrenes in soil. *Science* **1961**, *134*, 474–475. [CrossRef] [PubMed]
2. Ravindra, K.; Sokhi, R.; Van Grieken, R. Atmospheric polycyclic aromatic hydrocarbons: Source attribution, emission factors and regulation. *Atmos. Environ.* **2008**, *42*, 2895–2921. [CrossRef]
3. Shen, G.F.; Tao, S.; Wei, S.Y.; Chen, Y.C.; Zhang, Y.Y.; Shen, H.Z.; Huang, Y.; Zhu, D.; Yuan, C.Y.; Wang, H.C.; et al. Field measurement of emission factors of PM, EC, OC, parent; nitro-; and oxy-polycyclic aromatic hydrocarbons for residential briquette, coal cake; and wood in Rural Shanxi; China. *Environ. Sci. Technol.* **2013**, *47*, 2998–3005. [CrossRef] [PubMed]
4. Lafamme, R.E.; Hites, R.A. Global distribution of polycyclic aromatic hydrocarbons in recent sediments. *Geochim. Cosmochim. Acta* **1978**, *42*, 289–303. [CrossRef]
5. Masclet, P.; Hoyau, V.; Jaffrezo, J.L.; Cachier, H. Polycyclic aromatic hydrocarbon deposition on the ice sheet of Greenland. Part I: Superficial snow. *Atmos. Environ.* **2000**, *34*, 3195–3207. [CrossRef]
6. Cai, C.; Li, J.; Wu, D.; Wang, X.; Tsang, D.C.W.; Li, X.; Sun, J.; Zhu, L.; Shen, H.; Tao, S.; et al. Spatial distribution; emission source and health risk of parent PAHs and derivatives in surface soils from the Yangtze River Delta, eastern China. *Chemosphere* **2017**, *178*, 301–308. [CrossRef]
7. McCarrick, S.; Cunha, V.; Zapletal, O.; Vondráček, J.; Dreij, K. In vitro and in vivo genotoxicity of oxygenated polycyclic aromatic hydrocarbons. *Environ. Pollut.* **2019**, *246*, 678–687. [CrossRef]
8. Atkinson, R.; Arey, J. Atmospheric chemistry of gas-phase polycyclic aromatic hydrocarbons: Formation of atmospheric mutagens. *Environ. Health Perspect.* **1994**, *102*, 117–126.
9. Lundstedt, S.; White, P.A.; Lemieux, C.L.; Lynes, K.D.; Lambert, L.B.; Oberg, L.; Haglund, P.; Tysklind, M. Sources; fate; and toxic hazards of oxygenated polycyclic aromatic hydrocarbons (PAHs) at PAH-contaminated sites. *Ambio* **2007**, *36*, 475–485. [CrossRef]
10. Jung, D.J.K.; Klaus, T.; Fent, K. Cytochrome P450 induction by nitrated polycyclic aromatic hydrocarbons; azaarenes; and binary mixtures in fish hepatoma cell line PLHC-1. *Environ. Toxicol. Chem.* **2001**, *20*, 149–159. [CrossRef]
11. De Guidi, G.; Falciglia, P.P.; Catalfo, A.; De Guidi, G.; Fagone, S.; Vagliasindi, F.G.A. Soil contaminated with PAHs and nitro-PAHs: Contamination levels in an urban area of Catania (Sicily, southern Italy) and experimental results from simulated decontamination treatment. *Clean Technol. Environ. Policy* **2017**, *19*, 1121–1132. [CrossRef]
12. Musa Bandowe, B.A.; Wei, C.; Han, Y.; Cao, J.; Zhan, C.; Wilcke, W. Polycyclic aromatic compounds (PAHs, oxygenated PAHs, nitrated PAHs and azaarenes) in soils from China and their relationship with geographic location, land use and soil carbon fractions. *Sci. Total Environ.* **2019**, *690*, 1268–1276. [CrossRef] [PubMed]
13. Zhuo, S.; Du, W.; Shen, G.; Li, B.; Liu, J.; Cheng, H.; Xing, B.; Tao, S. Estimating relative contributions of primary and secondary sources of ambient nitrated and oxygenated polycyclic aromatic hydrocarbons. *Atmos. Environ.* **2017**, *159*, 126–134. [CrossRef]
14. Librando, V.; Fazzino, S. Quantification of polycyclic aromatic hydrocarbons and their nitro derivatives in atmospheric particulate matter of Augusta City. *Chemosphere* **1993**, *27*, 1649–1656. [CrossRef]
15. Hayakawa, K.; Tang, N.; Akutsu, K.; Murahashi, T.; Kakimoto, H.; Kizu, R. Comparison of polycyclic aromatic hydrocarbons and nitro polycyclic aromatic hydrocarbons in airborne particulates collected in downtown and suburban Kanazawa, Japan. *Atmos. Environ.* **2002**, *36*, 5535–5541. [CrossRef]
16. Albinet, A.; Leoz-Garziandia, E.; Budzinski, H.; Villenave, E. Polycyclic aromatic hydrocarbons (PAHs); nitrated PAHs and oxygenated PAHs in ambient air of the Marseilles area (South of France): Concentrations and sources. *Sci. Total Environ.* **2007**, *384*, 280–292. [CrossRef]
17. De Castro Vasconcellos, P.; Sanchez-Ccoyllo, O.; Balducci, C.; Mabilia, R.; Cecinato, A. Occurrence and concentration levels of nitro-PAH in the air of three Brazilian cities experiencing different emission impacts. *Water Air Soil Pollut.* **2008**, *190*, 87–94. [CrossRef]
18. Wild, S.R.; Jones, K.C. Polynuclear aromatic-hydrocarbons in the United-Kingdom environment—A preliminary source inventory and budget. *Environ. Pollut.* **1995**, *88*, 91–108. [CrossRef]
19. Bandowe, B.A.; Meusel, H. Nitrated polycyclic aromatic hydrocarbons (nitro-PAHs) in the environment—A review. *Sci. Total Environ.* **2017**, *581–582*, 237–257. [CrossRef]

20. Hoffmann, T.; Oelmann, Y.; Wilcke, W. Plant diversity enhances the natural attenuation of polycyclic aromatic compounds (PAHs and oxygenated PAHs) in grassland soils. *Soil Biol. Biochem.* **2019**, *129*, 60–70.
21. Burgos, W.D.; Pisutpaisal, N.; Mazzarese, M.C.; Chorover, J. Adsorption of quinolone to kaolinite and montmorillonite. *Environ. Eng. Sci.* **2002**, *19*, 59–64. [CrossRef]
22. Weigand, H.; Totsche, K.U.; Kögel-Knabner, I.; Annweiler, E.; Richnow, H.H.; Michaelis, W. Fate of anthracene in contaminated soil: Transport and biochemical transformation under unsaturated flow conditions. *Eur. J. Soil Sci.* **2002**, *53*, 71–81. [CrossRef]
23. Wilcke, W.; Kiesewetter, M.; Bandowe, B.A.M. Microbial formation and degradation of oxygen-containing polycyclic aromatic hydrocarbons (OPAHs) in soil during short-term incubation. *Environ. Pollut.* **2014**, *184*, 385–390. [CrossRef] [PubMed]
24. Schlanges, I.; Meyer, D.; Palm, W.U.; Ruck, W. Identification; quantification and distribution of PAC-metabolites, heterocyclic PAC and substituted PAC in groundwater samples of tar-contaminated sites from Germany. *Polycycl. Arom. Compd.* **2008**, *28*, 320–328. [CrossRef]
25. Bandowe, B.A.M.; Sobocka, J.; Wilcke, W. Oxygen-containing polycyclic aromatic hydrocarbons (OPAHs) in urban soils of Bratislava; Slovakia: Patterns; relation to PAHs and vertical distribution. *Environ. Pollut.* **2011**, *159*, 539–549. [CrossRef]
26. Atkinson, R.; Arey, J.; Zielinska, B.; Aschmann, S.M. Kinetics and nitro-products of the gas-phase OH and NO_3 radical-initiated reactions of naphthalene-d8; fluoranthene-d10, and pyrene. *Int. J. Chem. Kinet.* **1990**, *22*, 999–1014. [CrossRef]
27. Keyte, I.J.; Harrison, R.M.; Lammel, G. Chemical reactivity and long-range transport potential of polycyclic aromatic hydrocarbons—A review. *Chem. Soc. Rev.* **2013**, *42*, 9333–9391. [CrossRef]
28. Wilcke, W.; Amelung, W.; Martius, C.; Garcia, M.V.B.; Zech, W. Biological sources of polycyclic aromatic hydrocarbons (PAHs) in the Amazonian rain forest. *J. Plant Nutr. Soil Sci.* **2000**, *163*, 27–30. [CrossRef]
29. Moeckel, C.; Nizzetto, L.; Di Guardo, A.; Steinnes, E.; Freppaz, M.; Filippa, G.; Camporini, P.; Benner, J.; Jones, K.C. Persistent organic pollutants in boreal and montane soil profiles: Distribution; evidence of processes and implications for global cycling. *Environ. Sci. Technol.* **2008**, *42*, 8374–8380. [CrossRef]
30. Okere, U.V.; Semple, K.T. Biodegradation of PAHs in 'Pristine' Soils from different climatic regions. *J. Bioremed. Biodegrad.* **2012**, *S1*, 6. [CrossRef]
31. Huang, Q.; Weber, W.J. Interactions of soil-derived dissolved organic matter with phenol in peroxidase-catalyzed oxidative coupling reactions. *Environ. Sci. Technol.* **2004**, *38*, 338–344. [CrossRef] [PubMed]
32. Nam, J.J.; Thomas, G.O.; Jaward, F.M.; Steinnes, E.; Gustafsson, O.; Jones, K.C. PAHs in background soils from western Europe: Influence of atmospheric deposition and soil organic matter. *Chemosphere* **2008**, *70*, 1596–1602. [CrossRef] [PubMed]
33. Semple, K.T.; Morriss, A.W.J.; Paton, G.I. Bioavailability of hydrophobic organic contaminants in soils, fundamental concepts and techniques for analysis. *Eur. J. Soil Sci.* **2003**, *54*, 809–818. [CrossRef]
34. Wei, C.; Han, Y.; Bandowe, B.A.M.; Cao, J.; Huang, R.-J.; Ni, H.; Tian, J.; Wilcke, W. Occurrence, gas/particle partitioning and carcinogenic risk of polycyclic aromatic hydrocarbons and their oxygen- and nitrogen- containing derivatives in Xi'an, central China. *Sci. Total Environ.* **2015**, *505*, 814–822. [CrossRef] [PubMed]
35. Titaley, I.A.; Chlebowski, A.; Truong, L.; Tanguay, R.L.; Massey Simonich, S.L. Identification and toxicological evaluation of unsubstituted PAHs and novel PAH derivatives in pavement sealcoat products. *Environ. Sci. Technol. Lett.* **2016**, *3*, 234–242. [CrossRef]
36. Krauss, M.; Wilcke, W.; Zech, W. Polycyclic aromatic hydrocarbons and polychlorinated biphenyls in forest soils, depth distribution as indicator of different fate. *Environ. Pollut.* **2000**, *110*, 79–88. [CrossRef]
37. Arroyo, L.J.; Li, H.; Teppen, B.J.; Johnston, C.T.; Boyd, S.A. Oxidation of 1- naphthol coupled to reduction of structural $Fe3^{+}$ in smectite. *Clays Clay Miner.* **2005**, *53*, 587–596. [CrossRef]
38. Bamforth, S.M.; Singleton, I. Bioremediation of polycyclic aromatic hydrocarbons, current knowledge and future directions. *J. Chem. Technol. Biotechnol.* **2005**, *80*, 723–736. [CrossRef]
39. Lemieux, C.L.; Lambert, I.B.; Lundstedt, S.; Tysklind, M.; White, P.A. Mutagenic hazards of complex polycyclic aromatic hydrocarbon mixtures in contaminated soil. *Environ. Toxicol. Chem.* **2008**, *27*, 978–990. [CrossRef]
40. Murakami, M.; Yamada, J.; Kumata, H.; Takada, H. Sorptive behavior of nitro-PAHs in street runoff and their potential as indicators of diesel vehicle exhaust particles. *Environ. Sci. Technol.* **2008**, *42*, 1144–1150. [CrossRef]
41. Qiao, M.; Bai, Y.H.; Cao, W.; Huo, Y.; Zhao, X.; Liu, D.; Li, Z. Impact of secondary effluent from wastewater treatment plants on urban rivers, polycyclic aromatic hydrocarbons and derivatives. *Chemosphere* **2018**, *211*, 185–191. [CrossRef] [PubMed]
42. Wilcke, W.; Bandowe, B.A.M.; Lueso, M.G.; Ruppenthal, M.; del Valle, H.; Oelmann, Y. Polycyclic aromatic hydrocarbons (PAHs) and their polar derivatives (oxygenated PAHs; azaarenes) in soils along a climosequence in Argentina. *Sci. Total Environ.* **2014**, *473*, 317–325. [CrossRef] [PubMed]
43. Wei, C.; Bandowe, B.A.M.; Han, Y.; Cao, J.; Zhan, C.; Wilcke, W. Polycyclic aromatic hydrocarbons (PAHs) and their derivatives (alkyl-PAHs; oxygenated-PAHs; nitrated-PAHs and azaarenes) in urban road dusts from Xi'an; Central China. *Chemosphere* **2015**, *134*, 512–520. [CrossRef] [PubMed]
44. Bandowe, B.A.M.; Lueso, M.G.; Wilcke, W. Oxygenated polycyclic aromatic hydro- carbons and azaarenes in urban soils, a comparison of a tropical city (Bangkok) with two temperate cities (Bratislava and Gothenburg). *Chemosphere* **2014**, *107*, 407–414. [CrossRef] [PubMed]
45. Bandowe, B.A.M.; Nkansah, M.A. Occurrence; distribution and health risk from polycyclic aromatic compounds (PAHs; oxygenated-PAHs and azaarenes) in street dust from a major West African Metropolis. *Sci. Total Environ.* **2016**, *553*, 439–449. [CrossRef]

46. Brinkmann, M.; Maletz, S.; Krauss, M.; Bluhm, K.; Schiwy, S.; Kuckelkorn, J.; Tiehm, A.; Brack, W.; Hollert, H. Heterocyclic aromatic hydrocarbons show estrogenic activity upon metabolization in a recombinant transactivation assay. *Environ. Sci. Technol.* **2014**, *48*, 5892–5901. [CrossRef]
47. Chibwe, L.; Geier, M.C.; Nakamura, J.; Tanguay, R.L.; Aitken, M.D.; Simonich, S.L. Aerobic bioremediation of PAH contaminated soil results in increased genotoxicity and developmental toxicity. *Environ. Sci. Technol.* **2015**, *49*, 13889–13898. [CrossRef]
48. Zhang, B.-T.; Liu, H.; Liu, Y.; Teng, Y. Application Trends of Nanofibers in Analytical Chemistry. *TrAC Trends Anal. Chem. (Regul. Ed.)* **2020**, *131*, 115992. [CrossRef]
49. Qiao, M.; Fu, L.J.; Cao, W.; Bai, Y.H.; Huang, Q.X.; Zhao, X. Occurrence and removal of polycyclic aromatic hydrocarbons and their derivatives in an ecological wastewater treatment plant in South China and effluent impact to the receiving river. *Environ. Sci. Pollut. Res.* **2019**, *26*, 5638–5644. [CrossRef]
50. Han, Y.M.; Bandowe, B.A.M.; Wei, C.; Cao, J.J.; Wilcke, W.; Wang, G.H.; Ni, H.Y.; Jin, Z.D.; An, Z.S.; Yan, B.Z. Stronger association of polycyclic aromatic hydrocarbons with soot than with char in soils and sediments. *Chemosphere* **2015**, *119*, 1335–1345. [CrossRef]
51. Brorström-Lundén, E.; Remberger, M.; Kaj, L.; Hansson, K.; Palm-Cousins, A.; Andersson, H. *Results from the Swedish National Screening Programme 2008, Screening of Unintentionally Produced Organic Contaminants*; Report B1944; Swedish Environmental Research Institute (IVL): Göteborg, Sweden, 2010.
52. Sun, Z.; Zhu, Y.; Zhuo, S.; Liu, W.; Zeng, E.Y.; Wang, X.; Tao, S. Occurrence of nitro- and oxy-PAHs in agricultural soils in eastern China and excess lifetime cancer risks from human exposure through soil ingestion. *Environ. Int.* **2017**, *108*, 261–270. [CrossRef] [PubMed]
53. Yadav, I.C.; Devi, N.L.; Singh, V.K.; Li, J.; Zhang, G. Concentrations, sources and health risk of nitrated- and oxygenated-polycyclic aromatic hydrocarbon in urban indoor air and dust from four cities of Nepal. *Sci. Total Environ.* **2018**, *643*, 1013–1023. [CrossRef] [PubMed]
54. Idowu, O.; Semple, K.T.; Ramadass, K.; O'Connor, W.; Hansbro, P.; Thavaman, P. Analysis of polycyclic aromatic hydrocarbons (PAHs) and their polar derivatives in soils of an industrial heritage city of Australia. *Sci. Total Environ.* **2020**, *699*, 134303. [CrossRef] [PubMed]
55. Musa Bandowe, B.A.; Shukurov, N.; Kersten, M.; Wilcke, W. Polycyclic aromatic hydrocarbons (PAHs) and their oxygen-containing derivatives (OPAHs) in soils from the Angren industrial area, Uzbekistan. *Environ. Pollut.* **2010**, *158*, 2888–2899. [CrossRef] [PubMed]
56. Yadav, I.C.; Devi, N.L.; Li, J.; Zhang, G. Polycyclic aromatic hydrocarbons in house dust and surface soil in major urban regions of Nepal, implication on source apportionment and toxicological effect. *Sci. Total Environ.* **2018**, *616–617*, 223–235. [CrossRef] [PubMed]
57. Bleeker, E.; Wiegman, S.; de Voogt, P.; Kraak, M.; Leslie, H.A.; de Haas, E.; Admiraal, W. Toxicity of azaarenes. *Rev. Environ. Contam. Toxicol.* **2002**, *173*, 39–83.
58. Machala, M.; Ciganek, M.; Blaha, L.; Minksova, K.; Vondrak, J. Aryl hydrocarbon receptor-mediated and estrogenic activities of oxygenated polycyclic aromatic hydrocarbons and azaarenes originally identified in extracts of river sediments. *Environ. Toxicol. Chem.* **2001**, *20*, 2736–2743.
59. Xue, W.L.; Warshawsky, D. Metabolic activation of polycyclic and heterocyclic aromatic hydrocarbons and DNA damage, a review. *Toxicol. Appl. Pharmacol.* **2005**, *206*, 73–93. [CrossRef]
60. Wang, W.; Jariyasopit, N.; Schrlau, J.; Jia, Y.; Tao, S.; Yu, T.W.; Dashwood, R.H.; Zhang, W.; Wang, X.; Simonich, S.L. Concentration and photochemistry of PAHs; NPAHs; and OPAHs and toxicity of PM2.5 during the Beijing Olympic games. *Environ. Sci. Technol.* **2011**, *45*, 6887–6895. [CrossRef]
61. Andersson, J.T.; Achten, C. Time to say goodbye to the 16 EPA PAHs? Toward an up-to-date use of PACs for environmental purposes. *Polycycl. Arom. Compd.* **2015**, *35*, 330–354. [CrossRef]
62. Chlebowski, A.C.; Garcia, G.R.; La Du, J.K.; Bisson, W.H.; Truong, L.; Massey Simonich, S.L.; Tanguay, R.L. Mechanistic investigations into the developmental toxicity of nitrated and heterocyclic PAHs. *Toxicol. Sci.* **2017**, *157*, 246–259. [CrossRef] [PubMed]
63. Trine, L.S.D.; Davis, E.L.; Roper, C.; Truong, L.; Tanguay, R.L.; Simonich, S.L.M. Formation of PAH derivatives and increased developmental toxicity during steam enhanced extraction remediation of creosote contaminated superfund soil. *Environ. Sci. Technol.* **2019**, *53*, 4460–4469. [CrossRef] [PubMed]
64. WHO. *Selected Nitro-; and Nitro-Oxy-Polycyclic Aromatic Hydrocarbons: Environmental Health Criteria*; WHO: Geneva, Switzerland, 2003; Volume 229.
65. The Most Neglected Threat to Public Health in China is Toxic Soil. *Economist*, 13 June 2017.
66. Doucette, W.; Dettenmaier, E.K.; Bugbee, B.; Mackay, D. Mass transfer from soils to plants. In *Handbook of Chemical Mass Transport in the Environment*; Thobodeaux, L.J., Mackay, D., Eds.; CRS Press: Boca Raton, FL, USA, 2011; pp. 389–411.
67. Rengarajan, T.; Rajendran, P.; Nandakumar, N.; Lokeshkumar, B.; Rajendran, P.; Nishigaki, I. Exposure to polycyclic aromatic hydrocarbons with special focus on cancer. *Asian Pac. J. Trop. Biomed.* **2015**, *5*, 182–189. [CrossRef]
68. Yang, J.; Zhang, H.; Zhang, H.; Wang, W.; Liu, Y.; Fan, Y. Smoking modify the effects of polycyclic aromatic hydrocarbons exposure on oxidative damage to DNA in coke oven workers. *Int. Arch. Occup. Environ. Health* **2017**, *90*, 423–431. [CrossRef]

69. Mortamais, M.; Pujol, J.; van Drooge, B.L.; Macia, D.; Martinez-Vilavella, G.; Reynes, C.; Sabatier, R.; Rivas, I.; Grimalt, J.; Forns, J.; et al. Effect of exposure to polycyclic aromatic hydrocarbons on basal ganglia and attention-deficit hyperactivity disorder symptoms in primary school children. *Environ. Int.* **2017**, *105*, 12–19. [CrossRef]
70. Yaffe, D.; Cohen, Y.; Arey, J.; Grosovsky, A.J. Multimedia analysis of PAHs and nitro- PAH daughter products in the Los Angeles Basin. *Risk Anal.* **2001**, *21*, 275–294. [CrossRef]
71. Ciganek, M.; Raszyk, J.; Kohoutek, J.; Ansorgová, A.; Salava, J.; Palac, J. Polycyclic aromatic hydrocarbons (PAHs; nitro-PAHs; oxy-PAHs) polychlorinated biphenyls (PCBs) and organic chlorinated pesticides (OCPs) in the indoor and outdoor air of pig and cattle houses. *Vet. Med.* **2000**, *43*, 217–226.
72. Miller-Schulze, J.P.; Paulsen, M.; Toriba, A.; Tang, N.; Hayakawa, K.; Tamura, K.; Dong, L.; Zhang, X.; Simpson, C.D. Exposures to particulate air pollution and nitro-polycyclic aromatic hydrocarbons among taxi drivers in Shenyang; China. *Environ. Sci. Technol.* **2010**, *44*, 216–221. [CrossRef]
73. IARC. Diesel and gasoline engine exhausts and some nitroarenes. In *IARC Monographs on the Evaluation of Carcinogenic Risks to Humans*; IARC: Lyon, France, 2012; Volume 105, pp. 1–703.
74. Zwirner-Baier, I.; Neumann, H.G. Polycyclic nitroarenes (nitro-PAHs) as biomarkers of exposure to diesel exhaust. *Mutat. Res. Genet. Toxicol. Environ. Mutagen.* **1999**, *141*, 135–144. [CrossRef]
75. Seidel, A.; Dahmann, D.; Krekeler, H.; Jacob, J. Biomonitoring of polycyclic aromatic compounds in the urine of mining workers occupationally exposed to diesel exhaust. *Int. J. Hyg. Environ. Health* **2002**, *204*, 333–338. [CrossRef]
76. Alves, C.A.; Vicente, A.M.P.; Gomes, J.; Nunes, T.; Duarte, M.; Bandowe, B.A.M. Polycyclic aromatic hydrocarbons (PAHs) and their derivatives (oxygenated-PAHs, nitrated-PAHs and azaarenes) in size-fractionated particles emitted in an urban road tunnel. *Atmos. Res.* **2016**, *180*, 128–137. [CrossRef]
77. Baird, W.M.; Hooven, L.A.; Mahadevan, B. Carcinogenic polycyclic aromatic hydrocarbon-DNA adducts and mechanism of action. Environ. *Mol. Mutagen.* **2005**, *45*, 106–114. [CrossRef] [PubMed]
78. Chen, S.-C.; Liao, C.-M. Health risk assessment on human exposed to environmental polycyclic aromatic hydrocarbons pollution sources. *Sci. Total Environ.* **2006**, *366*, 112–123. [CrossRef] [PubMed]
79. Lemieux, C.L.; Long, A.S.; Lambert, I.B.; Lundstedt, S.; Tysklind, M.; White, P.A. Cancer risk assessment of polycyclic aromatic hydrocarbon contaminated soils determined using bioassay-derived levels of benzo[a]pyrene equivalents. *Environ. Sci. Technol.* **2015**, *49*, 1797–1805. [CrossRef]
80. Martins, M.; Costa, P.M.; Ferreira, A.M.; Costa, M.H. Comparative DNA damage and oxidative effects of carcinogenic and non-carcinogenic sediment-bound PAHs in the gills of a bivalve. *Aquat. Toxicol.* **2013**, *142–143*, 85–95. [CrossRef]
81. Moorthy, B.; Chu, C.; Carlin, D.J. Polycyclic aromatic hydrocarbons, from metabolism to lung cancer. *Toxicol. Sci.* **2015**, *145*, 5–15. [CrossRef]
82. Rosenkranz, H.; Mermelstein, R. The mutagenic and carcinogenic properties of nitrated polycyclic aromatic hydrocarbons. In *Nitrated Polycyclic Aromatic Hydrocarbons*; White, C.M., Ed.; Huethig: Heidelberg, Germany, 1985; pp. 267–297.
83. Idowu, O.; Semple, K.T.; Ramadass, K.; O'Connor, W.; Hansbro, P.; Thavamani, P. Beyond the obvious, environmental health implications of polar polycyclic aromatic hydrocarbons. *Environ. Int.* **2019**, *123*, 543–557. [CrossRef]
84. Takahashi, K.; Asanoma, M.; Yoshida, S.; Ning, G.; Mori, H.; Horibe, Y.; Watanabe, T.; Hirayama, T.; Nukaya, H.; Mizutani, T. Identification of 1;3;6-trinitropyrene as a major mutagen in organic extracts of surface soil from Nagoya city; Japan. *Genes Environ.* **2006**, *28*, 160–166. [CrossRef]
85. Watanabe, T.; Takahashi, K.; Konishi, E.; Hoshino, Y.; Hasei, T.; Asanoma, M.; Hirayama, T.; Wakabayashi, K. Mutagenicity of surface soil from residential areas in Kyoto city; Japan; and identification of major mutagens. *Mutat. Res.* **2008**, *649*, 201–212. [CrossRef]
86. Hasei, T.; Watanabe, T.; Endo, O.; Sugita, K.; Asanoma, M.; Goto, S.; Hirayama, T. Determination of 3;6-dinitrobenzo[e]pyrene in surface soil and airborne particles; and its possible sources; diesel particles and incinerator dusts. *J. Health Sci.* **2009**, *55*, 567–577. [CrossRef]
87. Kawanishi, M.; Watanabe, T.; Hagio, S.; Ogo, S.; Shimohara, C.; Jouchi, R.; Takayama, S.; Hasei, T.; Hirayama, T.; Oda, Y.; et al. Genotoxicity of 3;6- dinitrobenzo[e]pyrene; a novel mutagen in ambient air and surface soil; in mammalian cells in vitro and in vivo. *Mutagenesis* **2009**, *24*, 279–284. [CrossRef] [PubMed]
88. Kato, T.; Totsuka, Y.; Hasei, T.; Watanabe, T.; Wakabayashi, K.; Kinae, N.; Masuda, S. In vivo examination of the genotoxicity of the urban air and surface soil pollutant, 3;6-dinitrobenzo[e]pyrene; with intraperitoneal and intratracheal administration. *Environ. Toxicol.* **2013**, *28*, 588–594. [CrossRef] [PubMed]
89. IPCS. Selected Nitro- and Nitro-oxy-polycyclic Aromatic Hydrocarbons. Environ Health Criteria No. 229. In *International Programme on Chemical Safety*; IPCS: Geneva, Switzerland, 2003; pp. 1–400.
90. Collins, J.; Brown, J.; Alexeeff, G.; Salmon, A. Potency equivalency factors for some polycyclic aromatic hydrocarbons and polycyclic aromatic hydrocarbon derivatives. *Regul. Toxicol. Pharmacol.* **1998**, *28*, 45–54. [CrossRef]
91. Davie-Martin, C.L.; Stratton, K.G.; Teeguarden, J.G.; Waters, K.M.; Simonich, S.L.M. Implications of bioremediation of polycyclic aromatic hydrocarbon contaminated soils for human health and cancer risk. *Environ. Sci. Technol.* **2017**, *51*, 9458–9468. [CrossRef]
92. Pedersen, D.U.; Durant, J.L.; Penman, B.W.; Crespi, C.L.; Hemond, H.F.; Lafleur, A.L. Human-cell mutagens in respirable airborne particles in the northeastern United States. 1. Mutagenicity of fractionated samples. *Environ. Sci. Technol.* **2004**, *38*, 682–689. [CrossRef]

93. Bolton, J.L.; Trush, M.A.; Penning, T.M.; Dryhurst, G.; Monks, T.J. Role of quinones in toxicology. *Chem. Res. Toxicol.* **2000**, *13*, 135–160. [CrossRef] [PubMed]
94. Ghosal, D.; Ghosh, S.; Dutta, T.K.; Ahn, Y. Current state of knowledge in microbial degradation of polycyclic aromatic hydrocarbons (PAHs), a review. *Front. Microbiol.* **2016**, *7*, 1369.
95. Benigni, R.; Bossa, C. Mechanisms of chemical carcinogenicity and mutagenicity, a review with implications for predictive toxicology. *Chem. Rev.* **2011**, *111*, 2507–2536. [CrossRef] [PubMed]
96. Geier, M.C.; Chlebowski, A.C.; Truong, L.; Massey Simonich, S.L.; Anderson, K.A.; Tanguay, R.L. Comparative developmental toxicity of a comprehensive suite of polycyclic aromatic hydrocarbons. *Arch. Toxicol.* **2018**, *92*, 571–586. [CrossRef]
97. Wincent, E.; Jönsson, M.E.; Bottai, M.; Lundstedt, S.; Dreij, K. Aryl hydrocarbon receptor activation and developmental toxicity in zebrafish in response to soil extracts containing unsubstituted and oxygenated PAHs. *Environ. Sci. Technol.* **2015**, *49*, 3869–3877. [CrossRef]
98. Dranka, B.P.; Benavides, G.A.; Diers, A.R.; Giordano, S.; Zelickson, B.R.; Reily, C.; Zou, L.; Chatham, J.C.; Hill, B.G.; Zhang, J.; et al. Assessing bioenergetic function in response to oxidative stress by metabolic profiling. *Free Radic. Biol. Med.* **2011**, *51*, 1621–1635. [CrossRef] [PubMed]
99. Knecht, A.L.; Goodale, B.C.; Truong, L.; Simonich, M.T.; Swanson, A.J.; Matzke, M.M.; Anderson, K.A.; Waters, K.M.; Tanguay, R.L. Comparative developmental toxicity of environmentally relevant oxygenated PAHs. *Toxicol. Appl. Pharmacol.* **2013**, *271*, 266–275. [CrossRef] [PubMed]
100. Beije, B.; Moller, L. 2-Nitrofluorene and related compounds, prevalence and biological effects. *Mutat. Res.* **1988**, *196*, 177–209. [CrossRef] [PubMed]
101. El-Alawi, Y.S.; McConkey, B.J.; Dixon, D.G.; Greenberg, B.M. Measurement of short- and long-term toxicity of polycyclic aromatic hydrocarbons using luminescent bacteria. *Ecotoxicol. Environ. Saf.* **2002**, *51*, 12–21. [CrossRef]
102. Hu, J.; Nakamura, J.; Richardson, S.D.; Aitken, M.D. Evaluating the effects of bioremediation on genotoxicity of polycyclic aromatic hydrocarbon contaminated soil using genetically engineered; higher eukaryotic cell lines. *Environ. Sci. Technol.* **2012**, *46*, 4607–4613. [CrossRef]
103. Musa Bandowe, B.A.; Bigalke, M.; Kobza, J.; Wilcke, W. Sources and fate of polycyclic aromatic compounds (PAHs; oxygenated PAHs and azaarenes) in forest soil profiles opposite of an aluminium plant. *Sci. Total Environ.* **2018**, *630*, 83–95. [CrossRef]
104. Aranda, E.; Godoy, P.; Reina, R.; Badia-Fabregat, M.; Rosell, M.; Marco-Urrea, E.; García- Romera, I. Isolation of Ascomycota fungi with capability to transform PAHs, insights into the biodegradation mechanisms of Penicillium oxalicum. *Int. Biodeterior. Biodegrad.* **2017**, *122*, 141–150. [CrossRef]
105. Behera, B.K.; Das, A.; Sarkar, D.J.; Weerathunge, P.; Parida, P.K.; Das, B.K.; Thavamani, P.; Ramanathan, R.; Bansal, V. Polycyclic Aromatic Hydrocarbons (PAHs) in inland aquatic ecosystems, perils and remedies through biosensors and bioremediation. *Environ. Pollut.* **2018**, *241*, 212–233. [CrossRef]
106. Li, Y.; Liao, X.; Huling, S.G.; Xue, T.; Liu, Q.; Cao, H.; Lin, Q. The combined effects of surfactant solubilization and chemical oxidation on the removal of polycyclic aromatic hydrocarbon from soil. *Sci. Total Environ.* **2019**, *647*, 1106–1112. [CrossRef]
107. Keiluweit, M.; Kleber, M. Molecular-level interactions in soils and sediments: The role of aromatic π-systems. *Environ. Sci. Technol.* **2009**, *43*, 3421–3429. [CrossRef]
108. Geng, S.Y.; Xu, G.M.; Cao, W.; You, Y.; Zhu, Y.; Ding, A.Z.; Fan, F.Q.; Dou, J.F. Occurrence of polycyclic aromatic compounds and potentially toxic elements contamination and corresponding interdomain microbial community assembly in soil of an abandoned gas station. *Environ. Res.* **2022**, *212*, 113618. [CrossRef] [PubMed]
109. Bourhane, Z.; Lanzén, A.; Cagnon, C.; Said, O.B.; Mahmoudi, E.; Coulon, F.; Atai, E.; Borja, A.; Cravo-Laureau, C.; Duran, R. Microbial diversity alteration reveals biomarkers of contamination in soil-river-lake continuum. *J. Hazard Mater.* **2022**, *421*, 126789. [CrossRef] [PubMed]
110. Wang, B.B.; Teng, Y.; Yao, H.Y.; Christie, P. Detection of functional microorganisms in benzene [a] pyrene-contaminated soils using DNA-SIP technology. *J. Hazard Mater.* **2021**, *407*, 124788. [CrossRef] [PubMed]
111. Lu, C.; Hong, Y.; Odinga, E.S.; Liu, J.; Tsang, D.C.W.; Gao, Y.Z. Bacterial community and PAH-degrading genes in paddy soil and rice grain from PAH-contaminated area. *Appl. Soil Ecol.* **2021**, *158*, 103789. [CrossRef]
112. Chen, Z.S.; Hu, H.Y.; Xu, P.; Tang, H.Z. Soil bioremediation by Pseudomonas brassicacearum MPDS and its enzyme involved in degrading PAHs. *Sci. Total Environ.* **2022**, *813*, 152522. [CrossRef] [PubMed]
113. Liu, Y.L.; Hu, H.Y.; Zanaroli, G.; Xu, P.; Tang, H.Z. A Pseudomonas sp. strain uniquely degrades PAHs and heterocyclic derivatives via lateral dioxygenation pathways. *J. Hazard Mater.* **2021**, *403*, 123956. [CrossRef] [PubMed]
114. Brzeszcz, J.; Kaszycki, P. Aerobic bacteria degrading both n-alkanes and aromatic hydrocarbons: An undervalued strategy for metabolic diversity and flexibility. *Biodegradation* **2018**, *29*, 359–407. [CrossRef]

processes MDPI

Article

Incidence of Electric Field and Sulfuric Acid Concentration in Electrokinetic Remediation of Cobalt, Copper, and Nickel in Fresh Copper Mine Tailings

Rodrigo Ortiz-Soto [1,*], Daniela Leal [2], Claudia Gutierrez [2], Alvaro Aracena [1], Marcelo León [1], Andrea Lazo [2], Pamela Lazo [3], Lisbeth Ottosen [4] and Henrik Hansen [2]

[1] Escuela de Ingeniería Química, Pontificia Universidad Católica de Valparaíso, Avenida Brasil 2162, Valparaíso 2340025, Chile
[2] Departamento de Ingeniería Química y Ambiental, Universidad Técnica Federico Santa María, Avenida España 1680, Valparaíso 2390123, Chile
[3] Instituto de Química y Bioquímica, Facultad de Ciencias, Universidad de Valparaíso, Avenida Gran Bretaña 1111, Valparaíso 2360102, Chile
[4] Department of Civil Engineering, Technical University of Denmark, 2800 Kongens Lyngby, Denmark
* Correspondence: rodrigo.ortiz@pucv.cl; Tel.: +56-32-2372645

Abstract: In the present study, the assessment of heavy metal contaminant migration from fresh mine tailings was conducted using the electrokinetic remediation technique (EKR). In this sense, a pilot EKR cell was designed to evaluate the recovery potential of copper, nickel, and cobalt species. In particular, the focus was on the impacts of electric field intensity and pH in initial mixture and testing their interaction in copper, nickel, and cobalt migration. Experiments were made using a 2^2 factorial experimental design with a central point, using DC electric fields from 1.0 to 2.0 V cm^{-1} and H_2SO_4 pretreatment solutions from 1.0 to 2.0 mol L^{-1}, along with an ANOVA test with error reduction. The metal removal rates were approximately 7% for cobalt, neglectable for copper, and 6% for nickel. In the best cases, the highest concentrations by migration at the cathodic zone were 11%, 31%, and 30%, respectively. According to ANOVA tests, factor interaction was proven for each metal in the half cell near the cathode and the closest zone from the cathode specifically. Both factors affected metal concentrations, which indicates that when the goal aims for species accumulation in a narrower section, each factor has a significant effect, and their interaction makes a proven enhancement. Thus, using 2.0 V cm^{-1} and 2.0 mol L^{-1} showed a high improvement in metal concentration in the cathodic zone.

Keywords: electrokinetic remediation; heavy metal removal; mine tailing

1. Introduction

Heavy metals in soils are a complex, worldwide, urgent issue due to their toxicity and persistence [1]. Regarding mine tailings, given their magnitude and variety of heavy metals, their natural mechanisms of diffusion/advection allow them to be transported away from their confinement [2,3], risking various elements of ecosystems, such as water reservoirs [4], food [5], and the first link in the marine trophic chain [6], which directly impacts human health and the environment.

The mining industry has contributed immensely to the generation of heavy-metal-containing residues, including cobalt and nickel, which both have moderate toxic effects [7–9]. Therefore, every effort to remove these residues from the environment—including preventing their emanation, if possible—is urgent, especially when mining waste disposal methods are being revised to make mining a more sustainable industry [10,11]. To achieve removal of the residues, several technologies have been developed aiming for control and/or remediation, such as soil flushing, biological processes such as phytoremediation [12] and bioremediation [13], and electrochemically assisted processes, especially electrokinetic remediation (EKR) [14–16]. Although each technology has its benefits, soil flushing has problems due to the extent of extraction agents, and phytoremediation/bioremediation has complications in mine tailings because of a lack of nutrients. When phytoremediation and bioremediation are effective, they carry a potential risk of reintroducing these pollutants in the local trophic chain. On the other hand, the application of electricity allows the migration of ions contained in the liquid phase or available on the solid surface, regardless of how low their concentrations are. Therefore, this technology can transform mining residues into a potentially profitable asset regardless of energy consumption, since it can extract valuable inorganic species from depleted sources and can be applied in situ without the need to modify tailing dams. These advantages, in a global remediation market worth USD 30 billion [17], can make it more economically attractive. A few studies have been performed in mine tailings with this technology [18–24], also specifically in Chilean mine tailings [25–30], typically using a strong acid such as sulfuric acid to extract metals from the solid phase. The most fundamental factors in EKR are the electric field and acid concentration in the pretreatment solutions, even though this analysis has been conducted in one prior study [31], but it checked only one of the EKR sections.

For the aforementioned reasons, this study aims to establish the effect of the pH of the initial mixture (mine tailings with a 20% humidity by using an acid solution) and the electric field intensity applied between the electrodes in the metal migration yield in an EKR cell by identifying and assessing their interaction. A 2^2-factorial experimental design has been applied with a central point, in which the factors are the intensity of the electric field and the sulfuric acid concentration in the pretreatment solution, and an ANOVA test was performed in every partition of the cell, in addition to a MonteCarlo simulation for error reduction and migration difference verification between partitions and between experiments.

This statistical approach, which aims to establish more reliable conclusions with a better error propagation control, has never been applied in this way to fresh Chilean copper mine tailings for copper migration. Furthermore, cobalt and nickel EKR in this kind of solid have never registered before.

2. Materials and Methods

2.1. Tailing Samples and Reagents

Mine tailing samples were obtained from a copper processing plant in northern Chile, in the Punta del Cobre belt. According to del Real [32], copper ores from this region normally present copper as Chalcopyrite $CuFeS_2$, and the cobalt and nickel content normally appears in pyritic matrix by substitution $Fe(Co,Ni)S_2$. The sample was dried at 105 °C for several days until a constant weight was achieved, and the sample was pulverized in a ball mill to reduce particle size to below 500 microns, thus making it all a uniform size. Solid main characteristics are detailed in Table 1.

Table 1. Sample metal content, pH, and particle diameter.

Parameter	Value
pH	8.04 ± 0.06
Particle diameter, μm	212–350
Cu total concentration, mg kg^{-1}	486.50 ± 5.50
Co total concentration, mg kg^{-1}	89.10 ± 5.50
Ni total concentration, mg kg^{-1}	31.00 ± 1.40

H_2SO_4 (95%, analytic grade) was used to make different solutions used in this study: 5% (v/v) and 1.0, 1.5, and 2.0 mol L^{-1} each. Additionally, HNO_3 (65%, analytic grade) was used to form a 1:1 solution in volume, equivalent to 7.17 mol L^{-1}. $CaCl_2$ was used to form a 0.01 mol L^{-1} solution. HCl (37%, analytic grade) was needed without previous dilution. Distilled water was used when needed for every solution described.

2.2. Experimental Disposal and Measuring

As in Figure 1, an acrylic cell was used as a three-chamber container, separated by n° 131 filter paper. In the central chamber, a wet tailing (20% mass of H_2SO_4 solution to achieve solid saturation) sample was placed. In each extreme chamber, titanium electrodes and electrolytes were added from H_2SO_4 solutions, setting pH at 2 for the cathodic chamber and 4 for the anodic chamber. Each electrode was connected to an Extech 382,285 power supply and an Aktatum AMM-1139 multimeter. During operation, electrolytes were monitored and controlled during day hours via measurement of pH. To avoid hydroxide precipitation in the cathode, a few drops of sulfuric acid solution were added when needed. No pH control was possible during the night.

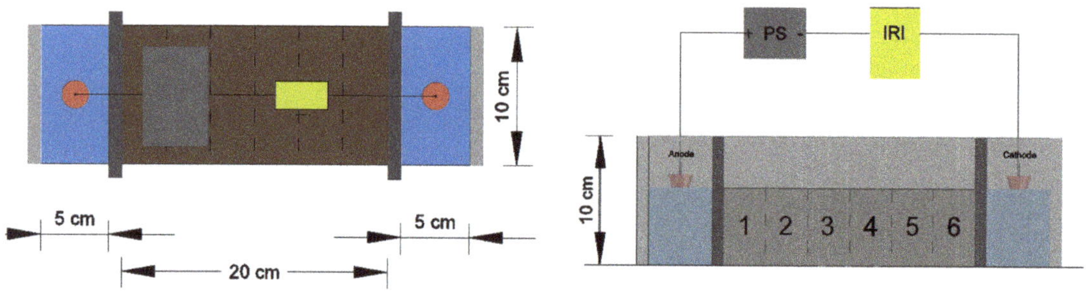

(**a**) (**b**)

Figure 1. EKR cell disposal. PS: power source. IRI: multimeter with data acquisition. (**a**) upper view; (**b**) front view. Numbers 1 through 6 are partitions used for measuring concentration results.

In each seven-day experiment, different sulfuric acid concentrations—the initial sample pH values of which varied from below 4.0 to 7.0, which allowed a different metal solubilization and electric field strength—were used. It was expected that a different metal migration extent for copper would result from varying the applied current; therefore, their current densities in five combinations are shown in Table 2. Effective operation time represents the time during which electrical current is applied through the cell before complete passivation. However, even with different effective times in some experiments, the electrical currents are in a similar range, and some experiments are comparable in these parameters. When the operation was completed, humid soil samples were taken from central chamber; divided into each one of six equal length partitions, as shown in Figure 1; numbered from 1 (nearest to anode) to 6 (nearest to cathode); and dried at 105 °C for 24 h. Next, total and soluble metal contents were measured in duplicate.

Table 2. Experimental design.

Experimental Number	Electric Field V cm^{-1}	H_2SO_4 Concentration mol L^{-1}	Sample pH at Beginning of Trial	Average Electrical Current mA	Effective Operation Time Days
1	1.0	1.0	6.93 ± 0.04	45.21	7.0
2	1.0	2.0	3.91 ± 0.02	157.48	7.0
3	2.0	1.0	7.04 ± 0.01	85.57	6.1
4	2.0	2.0	3.79 ± 0.03	339.63	7.0
5	1.5	1.5	4.51 ± 0.03	107.12	4.7

Total and soluble metal concentrations were determined for each partition and for the initial sample of mine tailing. In the case of total content, Danish Standard DS259:2003 was used by taking a 1 g sample of dried solid with 1:1 volume HNO_3 solution and placing the capped flask into the autoclave for 1 h at 200 kPa and 120 °C. Once the flask was cold, the solution was vacuum filtered with 0.45 μm filter paper, and distilled water was added to the filtered solution to make up 100 mL.

For the soluble content, 5 g of each sample was mixed with a 5% (v/v) H_2SO_4 solution for 30 min then vacuum filtered with 0.45 μm filter paper. Then, the filtered solution was mixed with 10 mL of HCl to form a 100 mL solution by adding distilled water. Liquid solutions in both cases were taken for measurement in Flame absorption spectrophotometry (AAS) in a Thermo Scientific M5 spectrometer.

2.3. Calculations

In EKR, a series of process performance indicators are usually required and are related to relevant species migration assessment throughout the entire cell, besides the need for statistical veracity conclusions. In this paper, the process is described in the following text.

2.3.1. Process Indicators

In this study, two indicators of metal ion movement were used, namely normalized metal concentration (κ) (Equation (1)) and global removal (Equation (2)).

$$\kappa_{e,I} = C_{e,i}/C_{e,initial} \quad (1)$$

$$\text{Total removal}_e = (m_{e,initial} - m_{e,final})/m_{e,initial} \cdot 100\% \quad (2)$$

where C is the concentration (in mg kg^{-1}) and m is the mass (in mg) throughout the whole cell. The subscript e indicates the species (Co, Cu, or Ni), and the subscript i indicates the partition considered (1 to 6), which are the solid sections of the cell central chamber in which each measurement was made after every EKR experiment, aiming for a description of their trajectory, as described in Figure 1.

2.3.2. Statistical Indicators

To ensure more solid conclusions, the statistical design of the experiments was utilized to determine the analyzed factors' effects: an Analysis of Variance [33] (ANOVA) was performed with 10% significance in every normal metal concentration (each metal and each partition, total and soluble), calculated as *p*-values (Equation (3)).

$$p - \text{Value} = P(F > f_{0,u,v}) = \int_{f_{0,u,v}}^{\infty} f(x)dx \quad (3)$$

where f(x) is Fisher's probability distribution with u degrees of freedom in the numerator and v degrees of freedom in the denominator. $f_{0,u,v}$ is the variance ratio between each analyzed factor and experimental error. When the *p*-value is lower than the significance level, the incidence of the factor in metal concentration in a particular partition is proven, with each factor as follows: electric field, acid concentration in pretreatment solution, and their interactions.

In order to establish significant differences in statistical terms, confidence intervals are needed. Since the sample data are reduced, statistically accurate conclusions are limited. Nonetheless, it is possible to mitigate its effect using the Monte Carlo method [34].

The sample standard deviation was calculated according to Equation (4) for each direct measurement.

$$s_{C_{m,i}} = \sqrt{\frac{\left(C_{m,i_1} - \overline{C}_{m,i}\right)^2 + \left(C_{m,i_2} - \overline{C}_{m,i}\right)^2}{n-1}} \quad (4)$$

where $s_{C_{m,i}}$ is the sample standard deviation for each metal m of normalized concentration in partition i (or initial), C_{m,i_1} and C_{m,i_2} are the measurements of each replicate, and $\overline{C}_{m,i}$ is the average concentration of both replicates, all in mg kg^{-1}; n is the sample size.

Then, 3000 sets of 100 numbers were generated with the generator from Microsoft Excel® using a normal distribution, where the population means and the standard deviations were their corresponding average concentrations and sample standard deviation in each case, respectively.

Later, each standard deviation of the normalized concentrations was determined from their indirect standard error, considering independence between samples (Equation (5)).

$$\alpha_{\kappa_{m,i,s,r}} = \sqrt{\left(\alpha_{C_{m,i,s,r}}/\overline{C}_{m,0,s,r}\right)^2 + \left(-\alpha_{C_{m,0,s,r}} \cdot \overline{C}_{m,i,s,r}/\overline{C}_{m,0,s,r}{}^2\right)^2} \quad (5)$$

where $\alpha_{\kappa_{m,\,i,s,r}}$ is the standard error for each metal m of normalized concentration in partition i, from experiment s. For the set r, $\alpha_{C_{m,\,i,s,r}}$ and $\alpha_{C_{m,0,s,r}}$ are the standard error of each metal of m concentration in partition i (or initial) for the experiment, respectively, from experiment s. Each one was calculated by dividing the sample standard deviation by the square root of 100. $\overline{C}_{m,i,s,r}$ and $\overline{C}_{m,0,s,r}$ are the average concentrations of each partition and initial in mg kg^{-1}.

Finally, the confidence interval, with a confidence level of 90%, is shown for each normalized concentration and difference between each possible pair (Equations (6) and (7)), respectively.

$$\kappa\prime_{m,i,s} = \frac{\sum_{r=1}^{3000}\left(\frac{\overline{C}_{m,i,s,r}}{\overline{C}_{m,0,s,r}} \pm t_{5\%,99} \cdot \alpha_{\kappa_{m,i,s,r}}\right)}{3000} = \overline{\kappa}_{m,i,s} \pm 1.660 \cdot \overline{\alpha}_{\kappa_{m,i,s}} \quad (6)$$

$$\kappa\prime_{m,i,s} - \kappa\prime_{m,j,b} = \overline{\kappa}_{m,i,s} - \overline{\kappa}_{m,j,b} \pm t_{5\%,DF} \cdot \sqrt{\overline{\alpha}_{\kappa_{m,i,s}}{}^2 + \overline{\alpha}_{\kappa_{m,j,b}}{}^2} \quad (7)$$

where $\kappa\prime_{m,\,i}$ and $\kappa\prime_{m,\,j}$ are the population means of normalized concentrations of metal *m* in partitions i and j for the experiments s and b, respectively. $\overline{\kappa}_{m,i,s}$ and $\overline{\kappa}_{m,j,b}$ are the average of the 3000 sets of normalized metal m in partitions i or j for experiments s and b, respectively. $\overline{\alpha}_{\kappa_{m,i,s}}$ and $\overline{\alpha}_{\kappa_{m,j,b}}$ are the average indirect standard errors. $t_{5\%,99}$ is the t-Student statistic parameter for bilateral confidence intervals for the chosen confidence level and 99 degrees of freedom. $t_{5\%,DF}$ is the same parameter type, but it is calculated for the confidence interval for differences between population means [33], considering equal variances between both treatments (Equation (8)).

$$DF = \left[\frac{\left(\overline{\alpha}_{\kappa_{m,i}}{}^2 + \overline{\alpha}_{\kappa_{m,j}}{}^2\right)^2}{\overline{\alpha}_{\kappa_{m,i}}{}^4 + \overline{\alpha}_{\kappa_{m,j}}{}^4} \cdot 99\right] \quad (8)$$

For concentration profiles, error bars are calculated from confidence intervals as Equation (6), and total removals are exhibited with their respective confidence intervals by the same method.

3. Results and Discussion

3.1. Cobalt

Indicators of cobalt total concentration achieved by experiments are shown in Figure 2. It was not possible to detect soluble concentrations of cobalt throughout the EKR cell, except for partition 6 in experiment 4, in which a concentration of 4.16 ± 0.30 mg kg^{-1} was observed. This phenomenon can be suggested from the work of del Real et al. [32] since the principal copper ore from the Punta del Cobre belt, where the sample is from, is Iron Oxide Copper-Gold (IOCG). In this kind of copper ore, cobalt is commonly present as sulfide, and in this mineral, the cobalt solubilization is extremely low, even in acidic environments [35].

After analyzing the results, cobalt migration was not observed in experiments 1, 2, and 5 since there are no statistically significant differences between any couple partitions. This was observed by noticing that all error bars for those experiments shared common values. Regarding experiment 3, partition 1 is significantly lower than any other partition near the cathode in the same experiment considering the error bars of partition 1 are lower than for partitions 4, 5, and 6. Between any other pair, this difference is not proven; thus, cobalt depletion was proven near the anode and migrated to every other partition equally. These results suggest that cobalt removal, at least from the anodic zone, is more influenced by voltage drop than acid concentration. In the case of experiment 4, partition 6 is significantly higher than any other partition in that experiment because its error bars are considerably higher than other partitions. Therefore, in this experiment, cobalt migration was observed seemingly from every cell sector to partition 6.

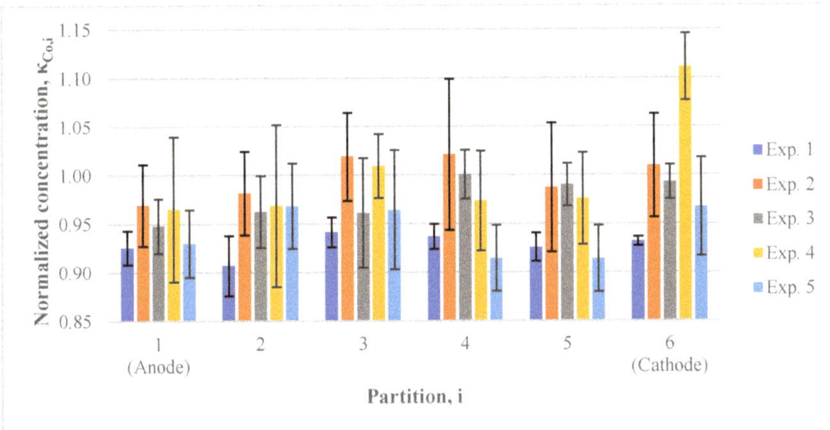

Figure 2. Total cobalt concentration profiles.

When comparing experiments, there is no difference in partition 1 for any experiment; therefore, cobalt removal in the anodic zone is indifferent to analyzed operating conditions. Regarding partition 6, results showed that experiment 4 raises cobalt concentration higher than any other experiment, and experiment 1 presents lower concentrations than experiments 2 and 3. This indicates that the most aggressive operation conditions are more effective in concentrating this species in the cathodic zone, reaching an 11% augmentation. On the other hand, minimally aggressive conditions are less effective in cobalt accumulation near the cathode than when a higher electric field or more acid initial solution is used. In addition, no difference was observed in this partition between experiments 2, 3, and 5. It can be suggested that their conditions are equivalent in the aim of concentrating cobalt near the cathode. From these results, it can be observed that cobalt migration is strongly promoted by using an electric field of 2.0 V cm^{-1} and a 2.0 mol L^{-1} acid solution, and this migration is not affected by anode proximity.

3.2. Copper

Figures 3 and 4 display total and soluble copper concentration profiles, respectively. Total copper migration from the anodic zone is concluded since the copper concentration in partition 1 is significantly lower than, at least, two other partitions in each experiment. As with cobalt, partition 6 of experiment 4 has the highest total copper concentration amongst every partition in any experiment, reaching a 31.00% augmentation, which suggests that factor interaction is effective for this ion. In experiments 3 and 5, a higher concentration is found in partition 3, and the total copper concentration is not significantly different between these experiments in any partition. This suggests that there is no impact from the change of electric field or acid concentration at these levels. Moreover, when acid concentration is 1.0 mol L^{-1}, migration comes uniquely from partition 1, but when a higher solution is used, migration can pass through the cell, being the central zone for experiments 3 and 5 or partition 6 in experiment 4.

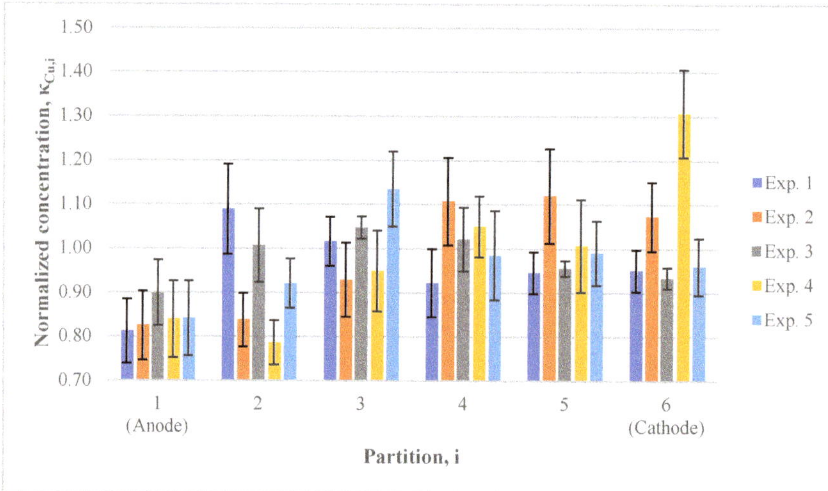

Figure 3. Total copper concentration.

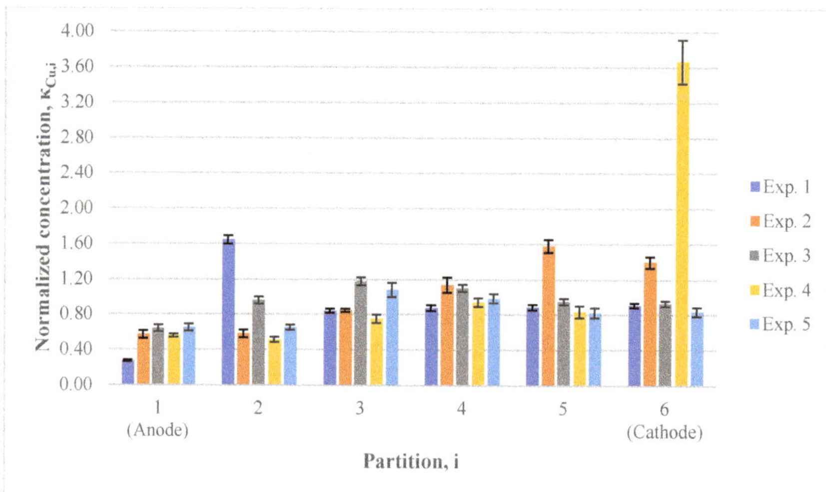

Figure 4. Soluble copper concentration.

Significant differences are proven throughout nearly the whole cell with regard to soluble copper, related to total copper profiles. Specifically, for partition 2 in experiment 1 and for partition 5 and 6 in experiment 2, soluble copper was 60% higher than initial concentrations, which suggests that, in the case of operating the EKR cell for more than seven days, copper migration to the cathodic zone could be significantly higher than observed here when the pretreatment solution was 1 mol L^{-1}. Nonetheless, in experiment 4, the weighted soluble copper proportion confidence interval throughout the cell was $32.95 \pm 4.03\%$, so copper solubilization is entirely evident in this case, since soluble copper is significantly higher than the initial concentration. This is supported by the understanding that IOCG ores present copper mainly as chalcopyrite, and for this mineralogical species, solubilization can happen when pH is between 2 and 4. There is an ORP higher than 0.3 V [36], which is the case, as was shown in a previous study [31] with the same ore and operating conditions. The pH of experiments 2 and 4 was from 2.0 to 4.0 in almost the whole cell, which shows the highest total copper concentrations near the cathode.

3.3. Nickel

Figure 5 shows total nickel concentration profiles. In each experiment, the lowest concentration was found in partition 1 because all results in this partition, including their error bars, are below 1.0. A comparison between the partitions of an experiment shows that they presented lower confidence values than other partitions. In the experiments 2, 3, and 5, higher total nickel concentrations were found in half of the cell near the cathode, corresponding to partitions 4, 5, and 6. For experiment 4, nickel removal is observed in the anodic zone since partitions 1 and 2 have significantly lower nickel concentrations than at least three other partitions in the central and cathodic zones. That removed nickel migrated to partitions 4 through 6, the latter of which showed the highest concentration amongst any partition and experiment, reaching a 30.00% augmentation. This indicates that electromigration is significantly higher in these operation conditions than in others in this study. Furthermore, for this case, differences between experiments—excluding experiment 4—are very similar, with their bigger difference being the zone in the cell in which the highest concentrations were observed. This suggests that, with the aim of nickel removal, higher values of acid concentration and/or electric field or any other reagents could help nickel solubilization.

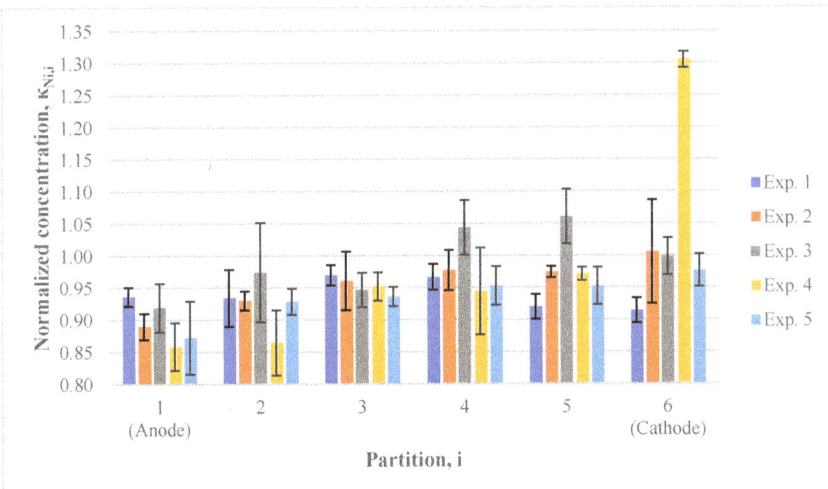

Figure 5. Total nickel concentration profiles.

With respect to soluble nickel, it was not possible to detect in most cases, including the original sample. Exceptions are partitions 2 and 3 in experiment 1, which had concentrations below 1.5 mg kg^{-1}, and in partition 6 from experiment 4, with 5.24 ± 0.52 mg kg^{-1}. An explanation of this solubilization can be sustained by the work of Lewis [35], who established that nickel sulfides can be dissolved mildly when pH is below 4.0, that is, below 0.1 mg L^{-1} in solution. Thus, in the case of experiment 1, anodic nickel could partially dissolve and migrate to the central zone of the cell. In the case of experiment 4, when partition 1 had a pH value around 2.5—when solubilization can be in the order of 1.0 mg L^{-1}, according to Lewis—and partitions 2 through 5 had pH values below 4.0, nickel leaching was considerably higher, and nickel migration accumulated in the cathodic zone.

3.4. Metal Removal

Table 3 displays global removals for each metal in any experiment. For cobalt and nickel, a minimal removal could be obtained in experiments 1 and 5 in both cases. Furthermore, experiment 2 presents removal for nickel only, and experiment 3 shows a very minimal cobalt removal, as removal is significantly lower than in experiments 1 and 5, since in all those cases removal is higher than the confidence limit. If the experiments are compared, best cobalt removal can be achieved using 1.0 V cm^{-1} of electric field and 1.0 mol L^{-1} of acid solution, since experiment 1 shows significantly higher results than experiment 5. In the case of nickel, there is not a significant difference between the same experiments 1 and 5 since the confidence intervals share values; therefore, best results are achieved in the same conditions as cobalt.

Table 3. Global removal for each metal in every experiment.

	Experiments				
	1	2	3	4	5
Total $removal_{Co}$	7.19% ± 0.74%	0.15% ± 4.49%	2.38% ± 2.09%	−0.57% ± 3.53%	5.40% ± 0.74%
Total $removal_{Cu}$	4.05% ± 5.22%	1.82% ± 5.60%	2.27% ± 2.46%	−0.32% ± 6.80%	2.57% ± 0.74%
Total $removal_{Ni}$	5.95% ± 1.65%	4.07% ± 2.40%	0.84% ± 3.08%	0.39% ± 1.81%	6.33% ± 0.74%

Nonetheless, copper removal was not observed in any experiment. These results are coherent with a previous similar study [25], in which a relatively a fresh copper sulphide tailing was also treated, and global removal was not observed. In contrast to other studies in which copper was removed in EKR experiments [37,38], the principal difference in this study is the chemical compound in which copper is found, since copper used in those other experiments was a highly soluble one, such as $CuSO_4$, which was added to the soil to form a synthetic sample. These findings, along with copper profiles, suggest solubilization can have a very important role in this kind of sample.

These indicators prove that, although experiment 4 shows better results in metal concentration, it is not the best combination to remove these metals from the cell. Even in the best conditions—experiment 1 for cobalt and nickel—nickel removal does not reach 2.0 mg kg^{-1} recovery, and cobalt obtains around 6.4 mg kg^{-1}. Nonetheless, these experimental runs were made for seven days. Thus, a potentially important removal could have been made if the experiment had been extended for more time, especially when each species was concentrated over 10% in the cathodic zone.

3.5. Interaction Proof

ANOVA *p*-value results for each metal are shown in Table 4. For cobalt, the effect of the electric field, pretreatment solution concentration, and their interaction is proven in partition 6. In partition 3, initial acidity has a relevant effect, while interaction between factors is proven in partitions 4 through 6. Nonetheless, factor effects in partitions 1 and 2 are not apparent.

For total copper, the electric field had a validated effect in partition 2, and sulfuric acid concentration had an incidence in each partition except partition 1. As with cobalt, factor interaction is proven in the same half-cell, and all factors and interactions are proven in partition 6. Furthermore, partition 1 shows no evidence of any factor influence. Regarding soluble copper, every factor and interaction influenced it throughout the entire cell, with the exception being partition 4, the factor interaction of which had an incidence in soluble copper concentration.

In the case of nickel, the initial acidity effect was shown throughout the cell, except in partition 3. Electric field incidence was proven in partitions 1, 5, and 6, and factor interaction was noted in partitions 2, 4, 5, and 6. As with every other species, each factor and its interaction incidences were shown in partition 6.

Table 4. Calculated p-values of each factor and their interaction for each partition.

Metal	Variability Source	Partition					
		1	2	3	4	5	6
Cobalt	Electric Field	60.70%	31.95%	71.70%	66.52%	13.35%	0.03%
	Initial Acidity	13.13%	10.09%	0.74%	16.93%	16.77%	0.02%
	Interaction	45.10%	14.54%	30.79%	3.09%	5.48%	4.80%
Total Copper	Electric Field	15.51%	6.95%	30.05%	48.71%	12.38%	0.33%
	Initial Acidity	46.24%	0.10%	1.33%	1.66%	1.33%	0.01%
	Interaction	27.77%	60.64%	81.29%	4.42%	8.02%	0.20%
Soluble Copper	Electric Field	0.02%	0.00%	0.52%	78.25%	0.01%	0.00%
	Initial Acidity	0.41%	0.00%	0.05%	15.86%	0.02%	0.00%
	Interaction	0.03%	0.00%	0.03%	0.09%	0.00%	0.00%
Nickel	Electric Field	8.37%	58.22%	20.50%	27.25%	0.08%	0.35%
	Initial Acidity	0.71%	5.27%	87.38%	6.20%	7.87%	0.03%
	Interaction	52.58%	6.60%	54.89%	3.40%	0.06%	0.04%

These results indicate that no matter which operating conditions are used, cobalt and copper are affected indifferently in the partition nearest to the anode, which suggests that proton formation in the anode and migration should be more effective in their solubilization and migration but in two opposite directions, such that copper solubilization is important, and cobalt has almost null solubilization. For nickel, the effect of acid could be explained by considerably higher solubilization with a lower pH in the conditions of these experiments.

Additionally, as in partition 6, every effect and interaction was proven; the inference is that, as metal migration goes directly to the cathode, every metal liberation and/or migration process, as mild as it could be, accumulates its effect in the partition near the cathode, even more so when metal removal was almost nonexistent in the experiments. Thus, if the operation aims to accumulate the species in a narrower section, each factor has a significant effect and its interaction as a declared enhancement.

According to ANOVA tests, factor interaction was proven by obtaining p-values below 10% in three partitions for each metal, being the closest to the cathode (partitions 4, 5, and 6). In partition 6 specifically, both main factors affected metal concentrations. In the case of cobalt and copper, in partition 1, neither the factor nor their interaction affected their total concentrations, which suggested that electrolysis in the anode is more determinant than the solubilization and/or migration of these metals in this zone. Regarding nickel, pretreatment solution concentration has proven effective in this partition since solubilization is considerably higher at the lower pH values of this experiment.

4. Conclusions

Although copper removal was not observed in any experiment, approximately 7.19% of cobalt and 5.95% of nickel were removed using 1.0 V cm^{-1} of electric field and 1.0 mol L^{-1} of pretreatment solution. Nonetheless, for each species, a concentration augmentation was achieved in partition 6, reaching 11.00% for cobalt, 31.00% for copper, and 30.00% for nickel, all of which were observed when the electric field was at 2.0 V cm^{-1} and the pretreatment solution was at 2.0 mol L^{-1}. It is highly recommended to repeat the experiments for, at least, twice the time used in this study since there is evidence of high metal concentration in the cathodic zone, which may be removed from the cell in that amount of time. Additionally, solubilization has a very important role in this kind of samples, and it is highly recommended to assess it separately from migration phenomena.

Regarding the ANOVA tests, interaction between the factors was proven in half of the cell near the cathode by having p-values below 10% in those partitions. Moreover, every factor and interaction for all species was observed in partition 6. Regarding nickel, pretreatment concentration reduces its concentration almost throughout the EKR cell. With regard to electric field, its incidence was proven in fewer partitions than initial acidity.

Therefore, a migration enhancement was obtained when electric field was at 2.0 V cm^{-1} and pretreatment solution was at 2.0 mol L^{-1}, with the aim of collecting more of each species near the cathode. A replication of the experiments with a longer operation time is strongly suggested to observe a better removal rate.

Author Contributions: Conceptualization, R.O.-S. and H.H.; methodology, C.G.; software, R.O.-S.; validation, A.A., H.H. and L.O.; formal analysis, R.O.-S.; investigation, D.L.; resources, A.L.; data curation, P.L.; writing—original draft preparation, R.O.-S. and M.L.; writing—review and editing, D.L. and M.L.; visualization, C.G.; supervision, H.H.; project administration, H.H.; funding acquisition, A.A. All authors have read and agreed to the published version of the manuscript.

Funding: This research received no external funding.

Institutional Review Board Statement: Not applicable.

Informed Consent Statement: Not applicable.

Data Availability Statement: Not applicable.

Conflicts of Interest: The authors declare no conflict of interest.

References

1. Lacasa, E.; Cotillas, S.; Saez, C.; Lobato, J.; Canizares, P.; Rodrigo, M.A. Environmental applications of electrochemical technology. What is needed to enable full-scale applications? *Curr. Opin. Electrochem.* **2019**, *16*, 149–156. [CrossRef]
2. Kelm, U.; Helle, S.; Matthies, R.; Morales, A. Distribution of trace elements in soils surrounding the El Teniente porphyry copper deposit, Chile: The influence of smelter emissions and a tailings deposit. *Environ. Geol.* **2009**, *57*, 365–376. [CrossRef]
3. Ramirez, M.; Massolo, S.; Frache, R.; Correa, J.A. Metal speciation and environmental impact in Sandy beaches due to El Salvador copper mine, Chile. *Mar. Pollut. Bull.* **2005**, *50*, 62–72. [CrossRef]
4. Fernández-Marchante, C.M.; Souza, F.L.; Millán, M.; Lobato, J.; Rodrigo, M.A. Can the green energies improve the sustainability of electrochemically-assisted soil remediation processes? *Sci. Total Environ.* **2022**, *803*, 149991–149999. [CrossRef] [PubMed]
5. Figueroa, A.; Cameselle, C.; Gouveia, S.; Hansen, H.K. Electrokinetic treatment of an agricultural soil contaminated with heavy metals. *J. Environ. Sci. Health A Tox. Hazard. Subst. Environ. Eng.* **2016**, *51*, 691–700. [CrossRef] [PubMed]
6. Correa, J.A.; Castilla, J.C.; Ramirez, M.; Varas, M.; Lagos, N.; Vergara, S.; Moenne, A.; Roman, D.; Brown, M.T. Copper, copper mine tailings and their effect on marine algae in northern Chile. *J. Appl. Phy̆siol.* **1999**, *11*, 57–67. [CrossRef]
7. Leyssens, L.; Vinck, B.; Van Der Straeten, C.; Wuyts, F.; Maes, L. Cobalt toxicity in humans—A review of the potential sources and systemic health effects. *Toxicology* **2017**, *387*, 43–56. [CrossRef] [PubMed]
8. Schaumlöffel, D. Nickel species: Analysis and toxic effects. *J. Trace Elem. Med. Biol.* **2012**, *26*, 1–6. [CrossRef]
9. Vardhan, K.H.; Kumar, P.S.; Panda, R.C. A review on heavy metal pollution, toxicity and remedial measures: Current trends and future perspectives. *J. Mol. Liq.* **2019**, *290*, 111197–111218. [CrossRef]
10. Araya, N.; Mamani Quiñonez, O.; Cisternas, L.A.; Kraslawski, A. Sustainable Development Goals in Mine Tailings Management: Targets and Indicators. *Mater. Proc.* **2021**, *5*, 82. [CrossRef]
11. Aznar-Sánchez, J.A.; García-Gómez, J.J.; Velasco-Muñoz, J.F.; Carretero-Gómez, A. Mining Waste and Its Sustainable Management: Advances in Worldwide Research. *Minerals* **2018**, *8*, 284. [CrossRef]
12. Qin, J.; Zhao, H.; Dai, M.; Zhao, P.; Chen, X.; Liu, H.; Lu, B. Speciation Distribution and Influencing Factors of Heavy Metals in Rhizosphere Soil of Miscanthus Floridulus in the Tailing Reservoir Area of Dabaoshan Iron Polymetallic Mine in Northern Guangdong. *Processes* **2022**, *10*, 1217. [CrossRef]
13. Tarfeen, N.; Nisa, K.U.; Hamid, B.; Bashir, Z.; Yatoo, A.M.; Dar, M.A.; Mohiddin, F.A.; Amin, Z.; Ahmad, R.A.; Sayyed, R.Z. Microbial Remediation: A Promising Tool for Reclamation of Contaminated Sites with Special Emphasis on Heavy Metal and Pesticide Pollution: A Review. *Processes* **2022**, *10*, 1358. [CrossRef]
14. Liu, L.; Li, W.; Song, W.; Guo, M. Remediation techniques for heavy metal-contaminated soils: Principles and applicability. *Sci. Total Environ.* **2018**, *633*, 206–219. [CrossRef]
15. Acar, Y.B.; Gale, R.J.; Alshawabkeh, A.N.; Marks, R.E.; Puppala, S.; Bricka, M.; Parker, R. Electrokinetic remediation: Basics and technology status. *J. Hazard. Mater.* **1995**, *40*, 117–137. [CrossRef]
16. Alshawabkeh, A.N. Electrokinetic soil remediation: Challenges and oportunitites. *Sep. Sci. Technol.* **2009**, *44*, 2171–2187. [CrossRef]
17. Singh, A.; Kuhad, R.C.; Ward, O.P. Biological Remediation of Soil: An Overview of Global Market and Available Technologies. In *Advances in Applied Bioremediation*; Soil Biology; Singh, A., Kuhad, R., Ward, O., Eds.; Springer: Berlin/Heidelberg, Germany, 2009; Volume 17, ISBN 978-3-540-89621-0.
18. Boonmeerati, U.; Sampanpanish, P. Enhancing Arsenic Phytoextraction of Dwarf Napier Grass (*Pennisetum purpureum* cv. Mott) from Gold Mine Tailings by Electrokinetics Remediation with Phosphate and EDTA. *J. Hazard. Toxic Radioact. Waste* **2021**, *25*, 04021027. [CrossRef]

19. Torabi, M.S.; Asadollahfardi, G.; Rezaee, M.; Panah, N.B. Electrokinetic Removal of Cd and Cu from Mine Tailing: EDTA Enhancement and Voltage Intensity Effects. *J. Hazard. Toxic Radioact. Waste* **2021**, *25*, 05020007. [CrossRef]
20. Asadollahfardi, G.; Sarmadi, M.S.; Rezaee, M.; Khodadadi-Darban, A.; Yazdani, M.; Paz-Garcia, J.M. Comparison if different extracting agents for the recovery of Pb and Zn through electrokinetic remediation of mine tailings. *J. Environ. Manag.* **2021**, *279*, 111728–111739. [CrossRef]
21. Boonmeerati, U.; Sampanpanish, P. Application of Phosphate and EDTA on As(V) Removal in Gold Mine Tailings by Electrokinetic Remediation. *EnvironmentAsia* **2020**, *13*, 1–12. [CrossRef]
22. Kirkelund, G.M.; Jensen, P.E.; Ottosen, L.M.; Pedersen, K.B. Comparison of two- and three-compartment cells for electrodialytic of heavy metals from contaminated material suspensions. *J. Hazard. Mater.* **2019**, *367*, 68–76. [CrossRef] [PubMed]
23. Rosa, M.A.; Egido, J.A.; Márquez, M.C. Enhanced electrochemical removal of arsenic and heavy metals from mine tailings. *J. Taiwan. Inst. Chem. E.* **2017**, *78*, 409–415. [CrossRef]
24. Demir, A.; Pamukcu, S.; Shrestha, R.A. Simultaneous Removal of Pb, Cd, and Zn from Heavily Contaminated Mine Tailing Soil Using Enhanced Electrochemical Process. *Environ. Eng. Sci.* **2015**, *32*, 416–424. [CrossRef]
25. Hansen, H.K.; Lamas, V.; Gutiérrez, C.; Nuñez, P.; Rojo, A.; Cameselle, C.; Ottosen, L.M. Electro-remediation of copper mine tailings. Comparing copper removal efficiencies for two tailings of different age. *Miner. Eng.* **2013**, *41*, 1–8. [CrossRef]
26. Hansen, H.K.; Rojo, A.; Ottosen, L.M. Electrokinetic remediation of copper mine tailings—Implementing bipolar electrodes. *Electrochim. Acta* **2007**, *52*, 3355–3359. [CrossRef]
27. Hansen, H.K.; Rojo, A. Testing pulsed electric fields in electroremediation of copper mine tailings. *Electrochim. Acta* **2007**, *52*, 3399–3405. [CrossRef]
28. Jensen, P.E.; Ottosen, L.M.; Hansen, H.K.; Bollwerk, S.; Belmonte, L.J.; Kirkelund, G.M. Suspended electrodialytic extraction of toxic elements for detoxification of three different mine tailings. *Int. J. Sustain. Dev. Plan.* **2016**, *11*, 119–127. [CrossRef]
29. Hansen, H.K.; Rojo, A.; Ottosen, L.M. Electrodialytic remediation of copper mine tailings. *J. Hazard. Mater.* **2005**, *117*, 179–183. [CrossRef]
30. Rojo, A.; Hansen, H.K.; Ottosen, L.M. Electrodialytic remediation of copper mine tailings: Comparing different operational conditions. *Miner. Eng.* **2006**, *19*, 500–504. [CrossRef]
31. Ortiz-Soto, R.; Leal, D.; Gutierrez, C.; Aracena, A.; Rojo, A.; Hansen, H.K. Electrokinetic remediation of manganese and zinc in copper mine tailings. *J. Hazard. Mater.* **2019**, *365*, 905–911. [CrossRef]
32. Del Real, I.; Thompson, J.F.H.; Simon, A.C.; Reich, M. Geochemical and Isotopic Signature of Pyrite as a Proxy for Fluid Source and Evolution in the Candelaria-Punta del Cobre Iron Oxide Copper-Gold District, Chile. *Econ. Geol.* **2020**, *115*, 1493–1518. [CrossRef]
33. Montgomery, D.C.; Runger, G.C. *Applied Statistics and Probability for Engineers*, 7th ed.; Wiley: Hoboken, NJ, USA, 2018; ISBN 978-1-119-40036-3.
34. Papadopoulos, C.E.; Yeung, H. Uncertainty estimation and Monte Carlo method. *Flow Meas. Instrum.* **2001**, *12*, 291–298. [CrossRef]
35. Lewis, A.E. Review of metal sulphide precipitation. *Hydrometallurgy* **2010**, *104*, 222–234. [CrossRef]
36. Córdoba, E.M.; Muñoz, J.A.; Blázquez, M.L.; González, F.; Ballester, A. Leaching of chalcopyrite with ferric ion. Part I: General aspects. *Hydrometallurgy* **2008**, *93*, 81–87. [CrossRef]
37. Chen, R.; Zhou, L.; Wang, W.; Cui, D.; Hao, D.; Guo, J. Enhanced Electrokinetic Remediation of Copper-Contaminated Soil by Combining Steel Slag and a Permeable Reactive Barrier. *Appl. Sci.* **2022**, *12*, 7981. [CrossRef]
38. Ge, X.; Xie, J.; Song, X.; Cao, X.; Wang, Y.; Xu, Z. Electrokinetic Remediation of Cadmiun (Cd), Copper (Cu) and Nickel (Ni) Co-contaminated soil with Oxalic Acid, Acetic Acid or Citric Acid as a Catholyte. *Int. J. Electrochem. Sci.* **2022**, *17*, 220444. [CrossRef]

processes

MDPI

Article

Effect of Rational Fertilizer for Eggplants on Nitrogen and Phosphorus Pollutants in Agricultural Water Bodies

Peiyuan Luo [1], Changchun Xin [1], Yuanhao Zhu [2], Yang Liu [3], Junhong Ling [1,4], Tianzhi Wang [1,*], Jun Huang [1,*] and Soon-Thiam Khu [1,2,5]

1 School of Environmental Science & Engineering, Tianjin University, Tianjin 300350, China
2 Zhejiang Tongji Vocational College of Science and Technology, Hangzhou 310000, China
3 School of Civil Engineering, Tianjin University, Tianjin 300072, China
4 Research Institute for Environmental Innovation, Tianjin 300457, China
5 Engineering Research Center of City Intelligence and Digital Governance, Ministry of Education of the People's Republic of China, Beijing 100083, China
* Correspondence: wangtianzhi@tju.edu.cn (T.W.); huangjun@tju.edu.cn (J.H.)

Abstract: Excessive fertilizer application in the cropping industry leads to excessive nitrogen and phosphorus in surrounding water bodies, which causes farmers to increase economic cost and damage the environment. To address the problem, we built a crop-soil runoff soil column test system, setting eight fertilizer application levels for eggplants. Then, crop yield, soil fertility and pollutant concentrations in the receiving water bodies were measured. The process of fertilizer application on the water quality of surrounding receiving water bodies and the rationality of fertilizer application were analyzed. The results showed that crop yield, soil fertility, and pollutant concentrations in the receiving water increased with the increase in fertilizer application. The crop yield stabilized when the fertilizer application amount was higher than 0.12 kg/m^2; the concentrations of total nitrogen (TN) and total phosphorus (TP) in the water increased significantly with the increase in fertilizer application, where particulate nitrogen (PN) and organic phosphorus (OP) were the key pollutants. In addition, crop growth had an interception effect on pollutant migration. The concentration of pollutants in the water was significantly lower in the late crop growth period (after flowering) than in the early. The crop had the best interception effect on pollutants when the fertilizer application was 0.12 kg/m^2. The concentrations of TN and TP in the water were 29.7% and 22.3% after the flowering period, being lower than those before the flowering period. Therefore, a reasonable value of 0.12 kg/m^2 is recommended for fertilizer application in this article. It can provide a reference for the fertilization system of eggplant planting in North China and theoretical support for the realization of clean production in small rural watershed planting.

Keywords: agricultural surface source pollution; nitrogen and phosphorus pollutant transport; agricultural receiving waters; excessive fertilizer application; clean production

1. Introduction

As a largely agricultural country, China is accelerating the shift in agriculture from an increased production-oriented system to a quality-oriented one now, realizing the transformation from a large agricultural country to a strong agricultural country. Therefore, the development of green and clean agriculture is of great significance to the construction of a strong agricultural country. However, with the overall progress of rural undertakings, there is the phenomenon of modern agriculture of exchanging environment for development, and the problem of surface source pollution in rural agriculture keeps emerging. The excess nitrogen and phosphorus in surrounding water bodies caused by excessive fertilizer application in the planting industry is a typical problem, with farmers both increasing economic costs and damaging the environment [1,2]. There is no study addressing the problems of high fertilizer input intensity in China's cultivated farmland area, nutrient loss due to higher than optimal fertilizer application, and serious pollution of the watershed water environment [3–5].

In response to these problems, many researchers have combined water conservation, surface source pollution control, and rural waste and sewage treatment [6,7] in the hope of achieving clean and non-eutrophic water bodies and a virtuous ecosystem cycle in small rural watersheds. However, the scale of small watersheds is equivalent to downscaling large watersheds, with the local amplification of conflicts and problems, prominent anisotropy, and increased heterogeneity [8]. For example, the peak point of pollutant content in runoff from small watersheds occurs closer to the peak point of rainfall and has a shorter response time to rainfall. Suspended matter, total nitrogen and total phosphorus showed higher linear correlation in runoff from small watersheds [9]. It causes the technical system of water pollution management in large watersheds to be not adaptable and poorly operable in small watersheds; water quality is more fragile in small watersheds; ecologically clean management measures in small watershed lack a system; water pollution is still serious.

It is the agricultural surface source pollution that is the main cause of the eutrophication of water bodies in small watersheds, contributing to more than 85% of water pollution in the watersheds [10]. As of 2015, the total fertilizer used in China was 5.98×10^{10} kg, accounting for 33% of the world's total fertilizer use, and the amount applied per unit area far exceeded the world average, where the utilization rate of nitrogen fertilizer was about 30–35%, while the utilization rate of phosphorus fertilizer was only 10–20% [11,12]. It means that the excessive use of chemical fertilizers will cause a large number of nutrients such as nitrogen and phosphorus to flow into water bodies, causing eutrophication. When pollutants enter rivers and lakes, even exceeding the self-purification capacity of water bodies, they will deteriorate the water quality and destroy the biological community groups of water bodies, and studies have shown that about 60% of the nitrogen and phosphorus content in surface polluted water bodies comes from chemical fertilizers [13]. For crops, fertilizer application over crop requirements and low fertilizer recovery rates in vegetable crops [14,15] may negatively affect crop yield and fruit size [16] and also lead to health risks (e.g., nitrate accumulation in leafy vegetables, micronutrient deficiencies in crops [17,18]). In conclusion, the problem of rational fertilizer application and control of agricultural surface pollution needs to be urgently addressed.

In recent years, researchers at home and abroad have agreed that four aspects are mainly included with the rational application of fertilizer: the amount of fertilizer applied, the period of fertilizer application, the method of fertilizer application, and the type of fertilizer (the "4R" technique, i.e., right amount, right time, right place, and right type). These four aspects are not isolated but are interrelated and affect each other. For example, the amount of fertilizer applied is determined first and foremost by the target yield, but also by the application method, period, and fertilizer variety [19]. If the latter three are unreasonable, it will contribute to the loss of a large number of nutrients during and after fertilization, which is not fully absorbed and utilized by the crop. To obtain a higher target yield, the farmer has to increase the amount of fertilizer applied to ensure that the crop absorbs enough nutrients. If the latter three tend to be reasonable, then the fertilizer applied can be fully absorbed and used by the crop, and there is no need to add additional fertilizer to "meet the losses". However, the current "4R" fertilization technology does not take into account the impact of fertilizer loss on the surrounding environment and does not form a reasonable value of fertilizer application based on the environmental damage effect of the surrounding water bodies, resulting in the loss of a large number of nutrients during and after fertilization, and the nutrients are not fully absorbed and utilized by the crop. In addition, relevant studies have shown that reducing the amount of fertilizer does not reduce vegetable yield [20,21], which shows that the determination of the reasonable value of fertilizer application has a great positive effect on increasing yield, protecting the environment and reducing costs, especially in areas with large planting areas; the determination of the reasonable value of fertilizer application will largely reduce farmers' fertilizer costs. For example, it has a year-round cash crop cultivation area of 1.19×10^8 m^2

with an annual output value of about USD 2.8 million in Tanchang County, Gansu Province, China. The fertilizer reduction space is huge.

Dimzon et al. [22] determined the water chemistry characteristics of water bodies through information on basic chemical indicators of water bodies (pH, redox potential, electric conductivity (EC), chemical oxygen demand (COD), and other parameters) and the content of substances in the water bodies. They studied regional water chemistry processes through correlations among the indicators. The method is well established and widely used for the study of urban water pollution processes, but the method has obvious limitations for plantation water pollution. Pollution from excessive fertilizer application in the plantation industry is characterized by strong dispersion, stochastic occurrence, and complex pollutant components [23], and current studies have focused on pollutant discharge [24,25], transport [26,27], and transformation processes [28]. Currently, the area of farmland at risk of agricultural surface source pollution in China exceeds 20 billion m^2. The contribution of nitrogen and phosphorus loss in the form of farm runoff to the nitrogen and phosphorus content of river water bodies exceeds 50%, respectively. Gan et al. [29] evaluated the situation of plantation surface source pollution in five counties around Chaohu Lake in Hefei City by monitoring the concentration of nitrogen and phosphorus loss from farmland. The average mass concentrations of TN and TP loss from cultivation in the region in the past three years were 3.48 mg/L and 0.602 mg/L, respectively, which were higher than the water quality standard of surface water category V. The export and pooling of nitrogen and phosphorus in rural sub-watersheds is closely related to the rainfall runoff production process, in which precipitation fully interacts with the soil and the nitrogen and phosphorus fertilizers applied by cultivation interact with rainwater, infiltrated water, and produced water components to influence the export and pooling of nitrogen and phosphorus pollutants to nearby water bodies. Yang et al. [30] monitored the nitrate–nitrogen concentration in rainfall runoff after fertilization and found that peak nitrate–nitrogen concentrations during runoff lagged behind peak confluent flow in surrounding water bodies, and nitrate–nitrogen concentrations in river water appeared to increase after the end of precipitation when runoff from surrounding water bodies gradually dropped back. Buchanan et al. [31] developed a phosphorus index model based on runoff travel time to improve the phosphorus index characterization of the effect of drainage on the potential risk of phosphorus loss. However, these studies could only qualitatively assess the key source areas of nitrogen and phosphorus loss from farmland at the watershed scale, and it was unclear to identify the key impact factors, which resulted in vague conclusions that could only provide the pollution sources that contributed more and could not provide the magnitude of the contribution of specific pollution sources to the water bodies, and could not quantify the pollution of water bodies in small watersheds due to excessive fertilizer application, which lacked practical guidance.

Based on the simulation results of different scenarios of the improved SWAT model, Wang et al. calculated and analyzed nitrate nitrogen and total phosphorus loads under each management scenario and their effects on crop yield, providing a theoretical basis for reducing non-point source pollution in the watershed and improving crop yield. However, this study is only suitable for a large watershed [32]. Li et al. analyzed the amount of fertilizer used in Wuhu City in 2017, calculated the amount of nitrogen and phosphorus lost in planting fields, and proposed problems such as unreasonable traditional fertilization method and fertilization structure, single fertilizer application type and large application amount, low penetration rate of efficient fertilization technology, and proposed and analyzed suitable fertilization technology, but did not propose a solution to the problem of fertilizer amount [33]. The researchers looked at the main factors contributing to the agricultural non-point source pollution and found that the pollution caused by fertilization was greater [34,35]. However, there is no recommendation on the appropriate level of fertilization. Wang et al. determined the temporal and spatial distribution of nitrate and ammonia nitrogen in the watershed by combining water sampling and numerical simulation and predicted the impact of agricultural management measures on reducing river nitrogen pollution [36], but the types of pollutants concerned were limited. Hou et al. established an improved output coefficient model based on rainfall and topography and studied the characteristics of pollution sources and non-point sources of TN and TP in Huangqianku Basin, which is an important drinking water source of the main tributaries of the lower Yellow River [37], and paid less attention to pollutant types [37]. In order to study the change of the content of each form of pollutants in water under different fertilization conditions in a small watershed, the migration pattern of nitrogen and phosphorus pollutants under the fertilization conditions of planting should be clarified, the mechanism of fertilization status on nitrogen and phosphorus pollutants in the receiving water bodies should be obtained, the key pollutants among the various forms of nitrogen and phosphorus pollutants should be found, reasonable values of fertilization in planting by considering nitrogen and phosphorus key pollutants, crop yield and soil fertility should be proposed. This article takes a small watershed as the research unit to carry out the study of the planting industry pollution prevention and control and selects eggplant in a typical small watershed within the Haihe River Basin in North China as the research object. We measured COD, eight nitrogen pollutants (total nitrogen (TN), particulate nitrogen (PN), dissolved total nitrogen (DTN), nitrate nitrogen (NO_3^--N), nitrite nitrogen (NO_2^--N), total Kjeldahl nitrogen (TKN), organic nitrogen (ON), ammonium nitrogen (NH_4^+-N)) and four phosphorus pollutants (total phosphorus (TP), phosphate (PO_4^{3-}), particulate phosphorus (PP), organic phosphorus (OP)) in agricultural runoff by building a crop-soil runoff water system and simulating the actual production and flow production process of the planting industry. The innovation of this article is that the key influencing factors are obtained from seven nitrogen pollutants and three phosphorus pollutants except TN and TP, and the contribution of specific pollution sources to the water body is given. In addition, the pollution caused by excessive fertilization to the water body in a small watershed is quantified. It is concluded that the crop can intercept the pollutants and the optimal fertilizer amount for eggplant planting is 0.12 kg/m^2. It can provide a reference for the fertilization system of eggplant cultivation in North China and provide theoretical support for realizing clean production in small rural watersheds for planting.

2. Materials and Methods

2.1. Crop-Soil Runoff Water Test System

The experiment was conducted in the hydraulic test hall of Tianjin University. The actual production and flow production process of the plantation was simulated by building a crop-soil runoff water system, as shown in Figure 1. The experimental setup used a large Plexiglas soil column to simulate the soil structure of agricultural planting. The soil was taken from the farmland soil of Xi Mengzhuang village section of the rural small watershed, Ji Canal watershed, within the Haihe watershed in North China, and the size of the Plexiglas was 1 m × 1 m × 0.5 m (to ensure the crop grows normally), and the area of the planting unit was 1 m × 1 m; the test crop was selected as eggplant with high water and fertilizer demand and high replanting index.

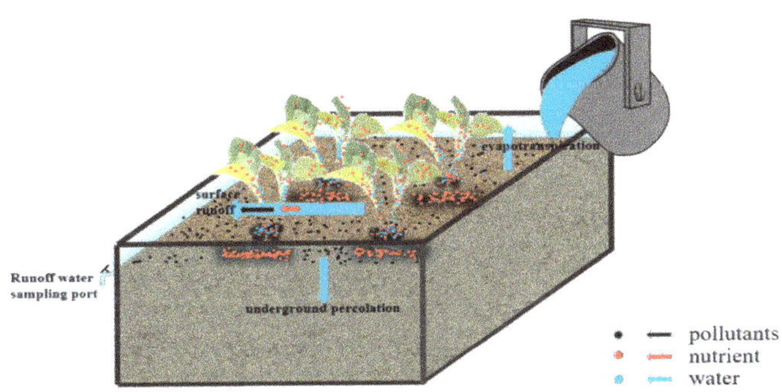

Figure 1. Schematic diagram of experimental equipment for crop-soil runoff water system.

Six fertilization patterns were set in this experiment, and the applied nitrogen fertilizer was calcium ammonium nitrate ($5Ca(NO_3)_2$-NH_4NO_3-$10H_2O$); phosphorus fertilizer was calcium dihydrogen phosphate ($Ca(H_2PO_4)_2$); and potassium fertilizer was potassium chloride (KCl). The fertilization amounts of the six fertilization patterns were shown in Table 1.

Table 1. Fertilization scheme for six plots.

Plot	$5Ca(NO_3)_2 \cdot NH_4NO_3 \cdot 10H_2O$ (kg/m^2)	$Ca(H_2PO_4)_2$ (kg/m^2)	KCl (kg/m^2)	Fertilizer Amount (kg/m^2)
1	0.012	0.009	0.009	0.03
2	0.024	0.018	0.018	0.06
3	0.036	0.027	0.027	0.09
4	0.048	0.036	0.036	0.12
5	0.060	0.045	0.045	0.15
6	0.072	0.054	0.054	0.18
CK	0	0	0	0

The system was irrigated and fertilized six times during the whole reproductive period, and irrigation was carried out 24 h after fertilization, and the amount of irrigation was set as shown in Table 2.

Table 2. Irrigation amount of eggplant at different growth stages (unit: mm).

Date	7.09	7.19	7.29	8.08	8.18	8.28
Growth Period	Planting	Slow seedling	Hardening of seedling	Flowering	Mature 1	Mature 2
Irrigation Amount	5	10	20	60	40	40
Reference	F-7.09	F-7.19	F-7.29	F-8.08	F-8.18	F-8.28

The eggplant was planted with sufficient bottom watering to keep the water in the permeable layer and to keep the soil moist for a long time with a slightly dry surface to prevent high humidity caused by frequent watering, which promoted seedling diseases. As the soil had little water-holding capacity, it was watered properly afterward so as not to allow seedlings to lose water. The seedlings were watered once when they were growing normally to ensure the expansion of the leaves and photosynthesis, thus producing some nutrients for their growth. When it starts to bear fruit, it was watered more frequently to ensure that the eggplants are large and have good color. The first fertilization was on 8 July 2022, planting was on 9 July and harvesting was on 1 September 2022. The specific fertilization dates are 8 July, 18 July, 28 July, 7 August, 17 August, and 27 August. Compound fertilizer was applied before planting and 2/3 of it was applied in general. Then it was plowed deeply and the other 1/3 of the fertilizer was applied and mixed with the soil. Planting is performed on a sunny day, with sufficient water in the furrow and the seedlings were planted after the water has seeped. The plants are spaced 30–35 cm apart and then mulched to a thickness level with the soil pile. Shallow tillage is carried out 3 days after planting to increase the ground temperature and promote the slowing down of seedlings. After the seedlings have slowed down, there was another mid-tillage and mulching, with the mid-tillage making a high monopoly of 12 to 15 cm, so that the monopoly surface was over the heap surface. Nitrogen and potassium fertilizer is applied in combination with watering time. When the eggplant sits in the fruit, then nitrogen fertilizer is applied. The specific experimental process is shown in Figure 2.

2.2. Test Content

2.2.1. Crop Yield

The harvesting period of eggplant in this experiment was from 10 September to 10 October 2022, and it was harvested a total of 4 times. Eggplant harvesting was measured by weighing the single fruit weight of eggplant with an electronic balance (accuracy 0.01 g).

Total output is the sum of four harvests per plot. Then, the number of single plant results was counted, and the average weight of the eggplant was calculated.

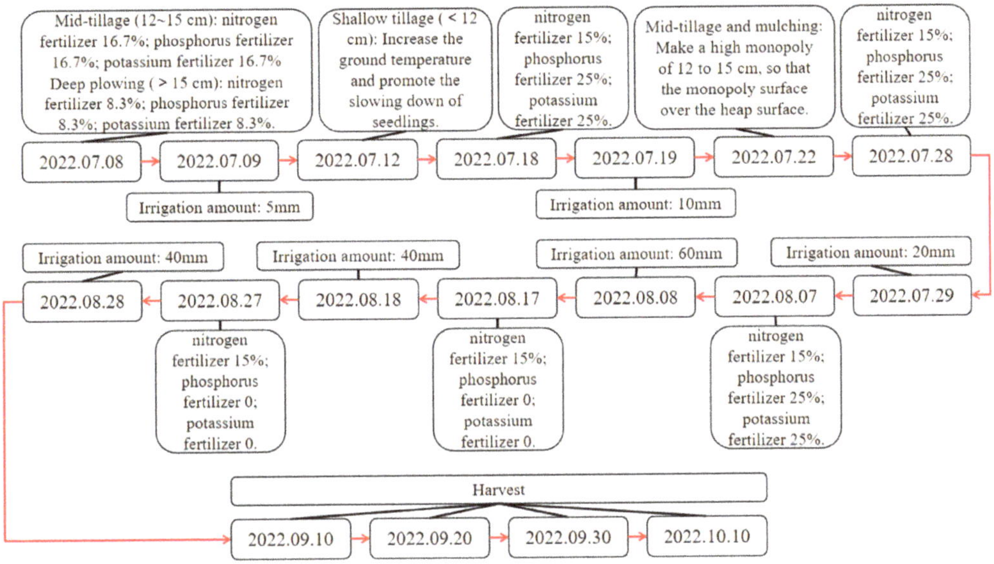

Figure 2. Specific experimental process.

2.2.2. Soil Fertility

Each plot was set up with 3 sampling points according to the principle of triangular uniform sampling, and each sampling point collected 3 surface soil samples to form a mixed soil sample. The soil samples were pretreated by air-drying, grinding, and sieving (<1 mm), and then tested for 5 indicators of soil nutrients, including available nitrogen, available phosphorus, available potassium, pH, and EC, using a soil tester (DK-TF1, Weifang, Shandong, China).

2.2.3. Contaminants in Water

After irrigation each time, runoff water samples were collected at the runoff water sampling port, and the collected samples were transferred into clean and dry sampling bottles, sealed with labels and numbers, and then transported back to the laboratory at a low temperature for the determination of different forms of nitrogen and phosphorus pollutants within 72 h. The indicators included COD, TN, TP, seven nitrogen pollutants (PN, DTN, NO_3^--N, NO_2^--N, TKN, ON and NH_4^+-N), and three kinds of phosphorus pollutants (PO_4^{3-}, PP and OP). The determination methods of pollutants are shown in Table 3.

2.2.4. Statistical Analysis

In this article, one plot was set for one fertilizer application; the content of available nitrogen, phosphorus, potassium, and the content of pollutants in water were taken three times and then the average value was obtained using Matlab-2019. The total production of eggplant was the sum of the four times harvests, and the single yield of eggplant was the ratio of total production of eggplant and the number of eggplants in each of the four times harvests. These data were plotted by Matlab-2019. In addition, the correlation diagram between pollutants and fertilizer amounts were manufactured by Origin-2022.

Table 3. The determination methods of pollutants.

Indicator	Method
TN	KH_2PO_4 oxidation—ultraviolet spectrophotometry
PN	The difference between TN and DTN
DTN	1. Filter with 0.4 μm filter hole 2. KH_2PO_4 oxidation—ultraviolet spectrophotometry
NO_3^--N	Phenol disulfonic acid photometry
NO_2^--N	N-1-Naphthylethylenediamine Dihydrochloride spectrophotometry
TKN	The difference between NO_3^--N and NO_2^--N
ON	The difference between TKN and NH_4^+-N
NH_4^+-N	Nessler's reagent spectrophotometry
TP	1. KH_2PO_4 digestion 2. Molybdenum-antimony resistance spectrophotometry
PO_4^{3-}	1. Filter with 0.4 μm filter hole 2. Mo-Sb Anti-spectrophotometer
PP	EDTA method
OP	1. Extracted by chloroform several times 2. Gas chromatography
COD	Potassium dichromate method

3. Results

3.1. Effect of Fertilizer Application on Soil Fertility in Plantation

The change in soil fertility with the increase in fertilizer application in different planting periods is shown in Figure 3.

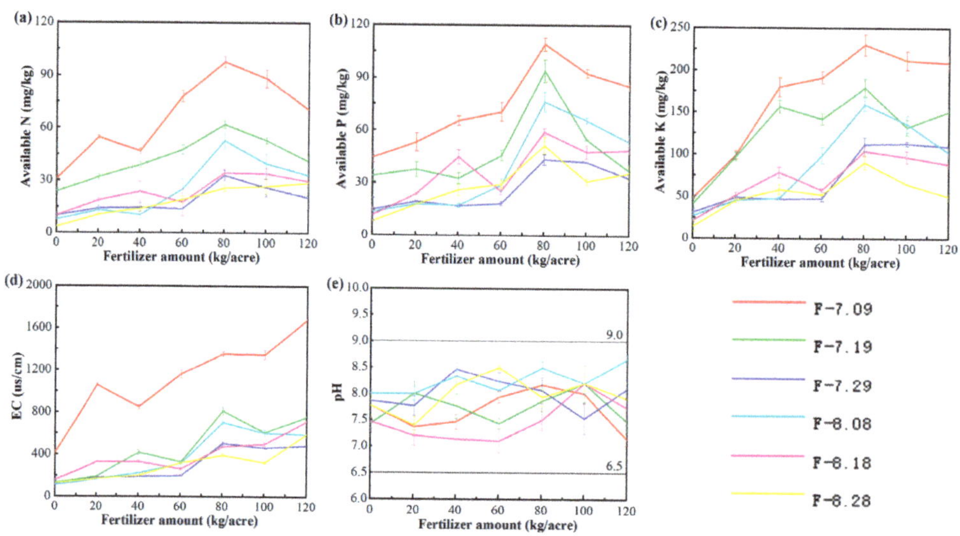

Figure 3. Relationship between soil fertility and fertilizer amount.

It can be seen that the highest content of available P in soil fertility was 96.5 mg/kg, which was 194.2% and 128.0% higher than the content of available N and available P, respectively. With the increase in fertilizer application, the content of available N, available P, and available K in the soil first increased and then decreased, and the highest content of available N, P, and K was found at 0.12 kg/m^2 with the average value of 50.9 mg/kg, 72.2 mg/kg, and 145.6 mg/kg, which were 36.9%, 48.3%, and 23.4% higher than that at 0.18 kg/m^2, respectively. Compared with the CK group (fertilizer application was 0), the available N, P, and K content increased 249.4%, 240.3%, and 378.9% at 0.12 kg/m^2, respectively. Soil EC showed a gradual increase with increasing fertilizer application, and the highest EC content was 804.6 μs/cm at 0.18 kg/m^2, which was 356.3% higher than that of the CK group. The variation in soil pH was independent of the fertilizer application amount and was in the range of 6.5 to 9.

In addition, when the fertilizer application was 0.12 kg/m^2, the available N, P, and K contents and EC generally showed a decreasing trend with the growth of the crop, but the flowering stage was more special, and the soil fertility at the flowering stage was higher than that at the hardening of the seedling and mature stages. The available N, P, and K contents and EC values at the flowering stage were 59.6%, 76.2%, 42.4%, and 39.6% higher than those at the hardening of the seedling, and 75.6%, 38.4%, 64.2%, and 62.9% higher than those at the mature stage, respectively.

3.2. Effect of Fertilizer Application on Crop Yield in Plantation

The relationship between eggplant yield and fertilizer application is shown in Figure 4.

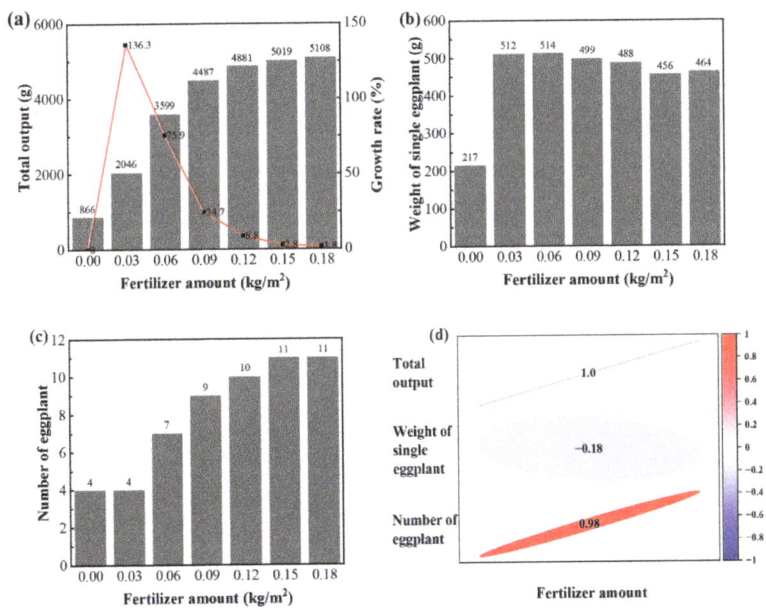

Figure 4. Relationship between fertilizer amount and total output, the weight of single eggplant, number of eggplants, and the correlation between them and fertilizer amount.

From Figure 4a, it can be seen that the total eggplant yield showed an increasing trend with the increase in fertilizer application, and the increase in eggplant yield showed a decreasing trend with the increase in fertilizer application. From Figure 4b, for individual eggplant weight, the individual eggplant weight of all fertilizer application groups was much higher than that of the CK group. When the fertilizer application amount was 0.03 kg/m^2~0.12 kg/m^2, the difference between individual eggplant weight was small and distributed in 488.0 g/unit~514.0 g/unit; when the fertilizer application amount was 0.12 kg/m^2~0.18 kg/m^2, the individual eggplant weight was significantly lower and the average weight was distributed from 464.0 g/unit to 488.0 g/unit; when the fertilizer application was 0.12 kg/m^2, the weight of individual eggplant was 488 g/unit, which was

5.1% higher than that when the fertilizer application was 0.18 kg/m^2. In addition, it can be seen from Figure 4c that the number of eggplants increased all the time with the increase in fertilizer application, with a minimum of four and a maximum of eleven. When the fertilizer application was 0.12 kg/m^2, the number of eggplants was 10, which was close to the optimal level. From Figure 4d, it can be seen that the correlation between the total output and the fertilizer amount was the highest (1) followed by the number of eggplants (0.98). The weight of single eggplant was negatively correlated with fertilizer amount and the value was small (−0.18).

3.3. The Effect of Fertilizer Application on Nitrogen and Phosphorus Pollutants in the Surrounding Water Bodies

The process of the influence of fertilizer application amount on pollutants in the receiving water bodies is shown in Figure 5.

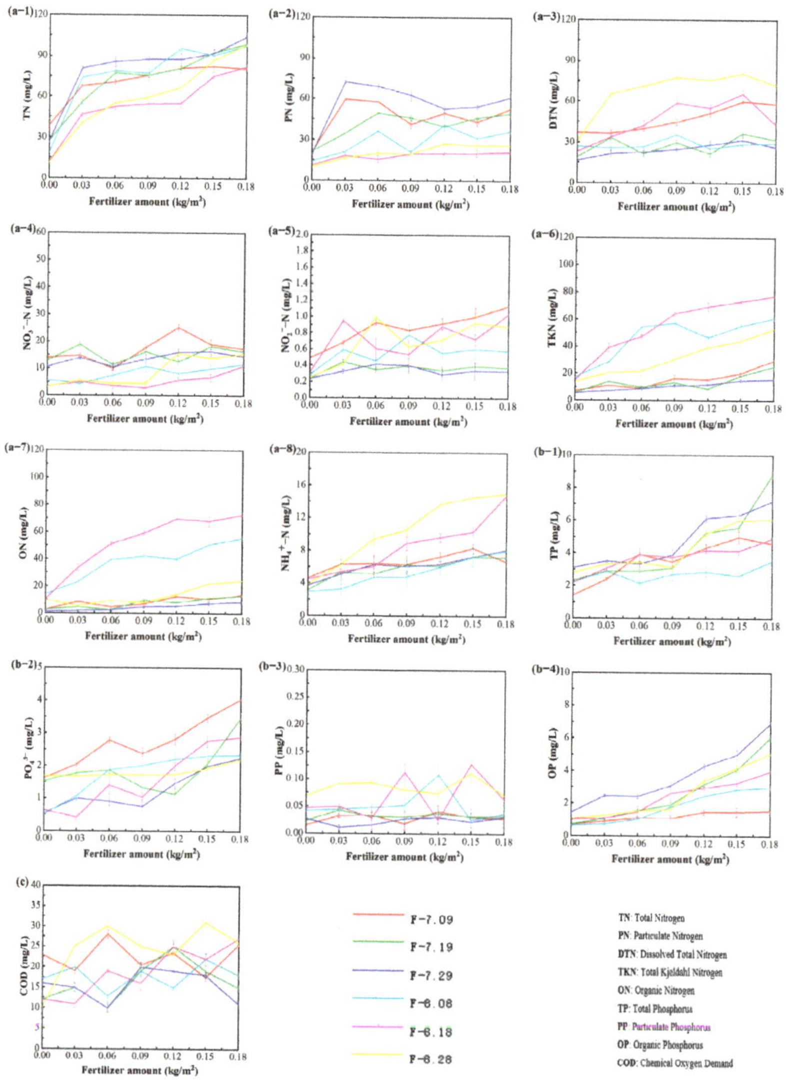

Figure 5. Relationship between fertilizer amount and pollutant concentration in runoff.

With the increase in fertilizer application amount, the TN and TP concentrations in the receiving water body showed an increasing trend, and at 0.18 kg/m^2, the TN and TP concentrations were the largest at 93.7 mg/L and 5.85 mg/L, which were 318.0% and 151.0% larger than the CK group, respectively, and the COD concentration did not change significantly at 15.0 mg/L~20.4 mg/L. For nitrogen pollutants, the order of pollutant concentrations in the water column was DN > PN > TKN > ON > NO_3^--N > NH_4^+-N > NO_2^--N except for TN, where the concentration of NO_2^--N was extremely low, only 1/100 of TN. Among the top four nitrogen pollutants with larger concentrations, TKN and ON were similar to TN in terms of concentration changes, and gradually increased with the increase in fertilizer application, and at 0.18 kg/m^2, the concentrations of TKN and DN were the largest, 43.4 mg/L and 43.3 mg/L, respectively, which were 303.2% and 70.0% larger than the CK group, respectively. Among these pollutants, OP and PO_4^{3-} were similar to TP and showed an increasing trend with the increase in fertilizer application, and the concentrations of OP and PO_4^{3-} were the largest at 0.18 kg/m^2, 4.6 mg/L and 2.9 mg/L, respectively. which were 390.2% and 128.2% greater than the CK group, respectively.

In addition, with the growth of the crop, the concentrations of PN and DN in the water fluctuated from planting to flowering, with PN decreasing and then increasing from flowering to maturity, and DN increasing from flowering to maturity; ON and TKN decreased from planting to hardening of the seedling, then increased until maturity one, and decreased again at maturity two. TN increased from planting to flowering, decreased at maturity one, and increased again at maturity two. TN increased from planting to flowering, decreased at maturity one, and increased again at maturity two. The TN and PN concentrations in water at maturity were 60.6 mg/L and 23.6 mg/L, respectively, which were 20.5% and 18.0% lower than those at flowering.

The trends of TP, OP, and PO_4^{3-} were fluctuating during the crop growth period, with TP and OP increasing significantly from the planting stage to the hardening of the seedling stage, decreasing at the flowering stage, and increasing again at the mature stage, and PO_4^{3-} conversely. The concentrations of TP and OP in the water at the flowering stage were 2.9 mg/L and 2.4 mg/L, respectively, which were 53.3% and 47.6% lower than those in the hardening of the seedling stage, indicating that the concentration of phosphorus pollutants in the receiving water changed more than that of nitrogen pollutants during the crop growth.

4. Discussion

The improvement in crop yield is first of all the result of crop variety renewal; after most varieties reach their required maximum yield fertilizer application, then increasing the fertilizer application will no longer have a yield increasing effect, and only new varieties with stronger fertilizer tolerance can ensure that the fertilizer application beyond the originally promoted varieties still has a yield increasing effect [38]. In addition, this experiment found that the increase in eggplant yield with increasing fertilizer application showed a decreasing trend with increasing fertilizer application. When the fertilizer application was higher than 0.12 kg/m^2, the weight of individual eggplants decreased significantly, which is consistent with the findings of Zhang et al. [39] and Xiao et al. [40]. They found that they were more sensitive to fertilizer tolerance, and excessive fertilizer application did not improve their yield and quality and even granularity. In this paper, we found that with the increase in fertilizer application, the content of available N, available P, and available K increased to the maximum when the fertilizer application was 0.12 kg/m^2 by analyzing the correlation between the eggplant yield and soil fertility. When the fertilizer application was higher than 0.12 kg/m^2, the content of available N and available P in the soil fertility decreased, and the content of nitrogen and phosphorus nutrients that could be absorbed by crops in the soil decreased. N is the main element of protein, nucleic acid, and phospholipid in living cells of crops, and is one of the constituent elements of chlorophyll, which has an important impact on crop photosynthesis. Phosphorus is beneficial to the smooth transition of plants from nutritional growth to reproductive growth, promoting the formation of reproductive organs and early flowering and fruiting; thus, nitrogen and phosphorus are essential for crop growth. Therefore, when the fertilizer application amount is higher than 0.12 kg/m^2, the eggplant yield no longer increases, and even the quality of the individual eggplants decreases.

In addition, the concentration of nitrogen and phosphorus pollutants in the receiving water bodies tended to increase with the increase in fertilizer application, which was due to the low utilization of nitrogen and phosphorus in the fertilizers used on the crops in this study, spreading fertilizers and the low nutrient utilization efficiency of vegetable crops because of their short growth time [41]. A large amount of nitrogen and phosphorus elements entered the receiving water bodies with runoff, causing water decent source pollution. In this paper, the relationships between eight nitrogen pollutants and four phosphorus pollutants were analyzed separately (Figure 6), and it was found that particulate nitrogen (PN) was the key nitrogen pollutant and organic phosphorus (OP) was the key phosphorus pollutant in the water. Nitrogen substances in fertilizers are transformed through decomposition by soil microorganisms to produce organic nitrogen, nitrate nitrogen, ammonium nitrogen, and particulate nitrogen. The inability of crops to utilize PN and the low utilization of DTN make the loss of PN and DTN the main forms of nitrogen loss from farmland [36]. DTN is mainly found in two forms: NH_4^+-N and NO_3^--N, and under aerobic conditions, NH_4^+-N is quickly oxidized to NO_3^--N. One part is absorbed by crops and the other part is adsorbed by soil colloids into the receiving water body with runoff, which makes the concentration of PN and DTN in the water higher, but the correlation between PN and TN is higher, which can be used as a key nitrogen pollutant in the receiving water body under different fertilizer application conditions. Phosphorus fertilizer application to soil is accompanied by a series of complex chemical and biological processes, including the mineralization of organic phosphorus and the biological fixation of inorganic phosphorus, the fixation of effective phosphorus, and the release process of insoluble phosphorus [42]. More phosphorus in the soil is released through the mineralization of organic phosphorus compounds, and then plants take up phosphorus from the soil solution as $H_2PO_4^-$ and HPO_4^{2-} with different efficiencies. The unmineralized organic phosphorus remains partly in the soil and partly excreted as runoff. It makes OP a key phosphorus contaminant in receiving waters.

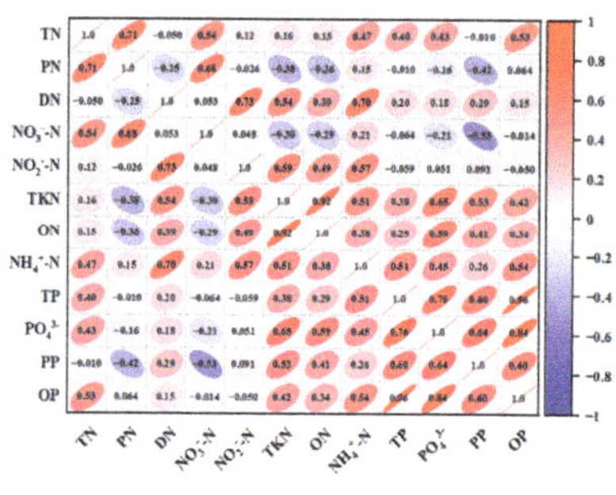

Figure 6. Correlation between different forms of nitrogen and phosphorus pollutants.

In addition, we found that during the flowering stage, the crop has the best interception effect on PN, TP, and OP, which can be reduced by 21.9%, 53.3%, and 47.6%, respectively, while during the maturity one, the interception effect on TN is the best, and the concentration can be reduced by 42.6%, indicating that crop growth will have a certain interception effect on the pollutant migration. The first point is due to the need for nutrients such as nitrogen and phosphorus for crop growth; the nitrogen absorbed by plants is mainly inorganic nitrogen, including ammonium nitrogen and nitrate nitrogen. Although low concentrations of nitrite can also be absorbed by plants, it is inherently absorbed in small amounts and high concentrations are harmful to plants and have no real nutritional value. Total nitrogen, total phosphorus, and organic phosphorus concentrations in water decrease the most during flowering and maturity because of increased root uptake and nutrient utilization capacity in late plant growth; Nendel et al. [43] found that in dry or temperate environments, potentially mineralized nitrogen accounts for a higher proportion of total soil nitrogen, and crops use fewer nutrients from fertilizer, the available N, P, and K originally contained in the soil are preferentially absorbed and used by crops resulting in larger concentrations of TN, TP and OP in runoff during the early stages of crop growth. In addition, soil fertility varies under different soil types. The soil used in the experiment is loam, which has good tillability, has the advantages of sandy soil and heavy loam, and has significantly higher soil fertility than heavy loam, which is more suitable for crop growth and development [44]. Chen et al. found that the comprehensive fertility of different soil types was yellow soil > paddy soil > lime soil > purple soil when exploring the relationship between soil types and soil fertility in a citrus orchard [45], which was also the reason why there were more available N, P and K in crop utilization soil. The number of microorganisms in the soil also affects the soil fertility. Chen et al. took tea trees of different planting years as research objects and found that there was a correlation between the number of microorganisms in rhizosphere soil and some soil fertility indexes [46]. The number of soil bacteria was significantly correlated with the soil organic matter and total phosphorus; the number of fungi was not significantly correlated with the soil nutrients; the number of actinomyces was significantly correlated with soil nutrients; the number of soil microorganisms reflected the level of soil fertility at a certain level [47]. Therefore, there may be more microorganisms in the soil selected in the experiment, and the soil fertility is higher, which provides part of the nutrients for the early growth of crops. In addition, irrigation water may also contain significant amounts of N. For example, a typical seasonal irrigation volume for sweet pepper is approximately 0.3 $m^3 \cdot m^{-2}$ with a nitrate concentration of 50 $mg \cdot L^{-1}$, corresponding to a total N input of 0.0034 $kg \cdot m^{-2}$ [48], and in this study crop growth could directly utilize inorganic nitrogen in water and utilize less nitrogen in the fertilizer, resulting in a larger total nitrogen concentration in the water at the beginning of crop growth and not showing a greater interception effect until maturity one. Second, we reduced irrigation from 60 mm to 40 mm at maturity and reduced irrigation, directly reducing the amount of leachate, which then promotes root growth [49–51], it may also increase plant uptake of water and nutrients [52] and reduce N and P leaching. On the other hand, reduced irrigation harms soil microbial load, microbial diversity, and enzyme activity [53], thereby inhibiting the conversion of N and P, which also helps to reduce their leaching. The second reason why crop growth would have an interception effect on pollutant transport is that crop roots precipitate particulate matter in water, and in this study, the degree of PN interception by the crop was greatest at flowering. Yang et al. [54] showed that plants not only remove some of the N and P from water through uptake but also precipitate the particulate matter by reducing the flow rate due to the well-developed root system of the crop and the substrates.

Finally, this paper discusses reasonable fertilization in the process of eggplant planting by combining the interrelationship between fertilization, the amount of soil, the crop fertility, and the environmental effects on the yield. When the fertilizer application amount was 0.12 kg/m^2, the soil fertility was better, and the available N, P, and K increased by 249.4%, 240.3%, and 378.9% compared with the CK group; the soil EC value can play a guiding role in plant fertilization and can reflect the actual condition of the soil salinity, which affects the transformation, existence state, and effectiveness of the soil nutrients and is the threshold value limiting plant and microbial activity. Relevant studies have shown that the optimal range of soil EC required for eggplant cultivation is 500 μs/cm to 750 μs/cm. In this study, when the fertilizer application rate was 0.12 kg/m^2, the soil EC was 710 μs/cm, which met the requirements for eggplant cultivation. Considering crop yield, when the fertilizer application amount was less than 0.12 kg/m^2, the increase in fertilizer application could promote crop growth, and when the fertilizer application amount was greater than or equal to 0.12 kg/m^2, the total yield remained the same, indicating that the fertilizer application amount was no longer a key factor limiting crop yield, but when the fertilizer application amount was higher than 0.12 kg/m^2, the yield of the individual eggplants would decrease. Considering environmental pollution, with the increase in fertilizer application, the concentration of pollutants also increases, but the main nitrogen and phosphorus pollutants all slow down the growth and tend to balance when the fertilizer application is 0.12 kg/m^2; when the fertilizer application is higher than 0.12 kg/m^2, the concentration of total nitrogen and phosphate increases significantly, which will pollute the water body to a great extent and bring pressure to the surrounding environment. Therefore, after considering the soil fertility, crop yield, and environmental pressure, the reasonable value of fertilizer application for eggplant planting is 0.12 kg/m^2.

The limitation of this study is that the influence of soil microorganism quantity on the soil's available nitrogen, phosphorus, and potassium for eggplant growth was not considered when exploring the influence of fertilizer amount on the soil's available nitrogen, phosphorus, and potassium. Our research group is designing related experiments to study the effect of soil microorganism quantity on the soil's available nitrogen, phosphorus, and potassium content, and further improve the part of the study on the effect of fertilizer application on available nitrogen, phosphorus, and potassium content.

5. Conclusions

This article takes eggplant in a typical small watershed within the Haihe River Basin in North China as the research object; we study the changes in crop growth and the content of various forms of pollutants in the receiving water under different fertilizer application amounts by building a crop-soil runoff water system and obtain the following conclusions.

(1) The key nitrogen and phosphorus pollutants are PN and OP, respectively. The average proportion of PN to TN is 50.9% and the average proportion of OP to TP is 60.6%.
(2) Crop growth will have a certain interception effect on pollutant migration. At the flowering stage, PN, TP, and OP were intercepted best by the crop, which can be reduced by 21.9%, 53.3%, and 47.6%, respectively, while TN was intercepted best at maturity one, and the concentration could be reduced by 42.6%.
(3) Combining the interrelationship between fertilizer application on soil, crop fertility, and environmental effects on yield in the process of eggplant planting, the reasonable value of fertilizer application for eggplant planting in this paper is suggested to be 0.12 kg/m^2. At this time, the soil fertility is the highest, the yield is close to the maximum, and the crop has the best interception effect on pollutants.

Author Contributions: Conceptualization, Established the research line and supervised the current work, T.W.; Performed the experiments and Data curation, C.X.; Writing-Reviewing and Editing, P.L.; Guide data analysis, Y.Z. and Y.L.; Guide drawing, J.L. and J.H.; Visualization, Supervision, S.-T.K. All authors have read and agreed to the published version of the manuscript.

Funding: This research was funded by the National Natural Science Fund Independent Innovation Fund of Tianjin University [2022XSU-0030, 2022XZ-0019]. And The APC was funded by J.H.

Institutional Review Board Statement: Not applicable.

Data Availability Statement: The data that support the findings of this study are available from the corresponding author upon reasonable request.

Acknowledgments: We are grateful for the financial support from the National Natural Science Fund Independent Innovation Fund of Tianjin University (2022XSU-0030, 2022XZ-0019). All individuals included in this section have consented to the acknowledgement.

Conflicts of Interest: The authors declare no conflict of interest.

References

1. Xiong, H.F. The effect of agricultural non-point Source Pollution of nitrogen and phosphorous on Lake Eutrophication. *IOP Conf. Ser. Earth Environ. Sci.* **2017**, *64*, 012061. [CrossRef]
2. Shou, C.G.; Du, H.S.; Liu, X.P. Research progress of source and mechanism of agricultural non-point source pollution in China. *Appl. Ecol. Environ. Res.* **2019**, *17*, 10611–10621. [CrossRef]

3. He, R.; Shao, C.F.; Shi, R.G.; Zhang, Z.Y.; Zhao, R. Development Trend and Driving Factors of Agricultural Chemical Fertilizer Efficiency in China. *Sustainability* **2020**, *12*, 4607. [CrossRef]
4. Thorburn, P.J.; Biggs, J.S.; Palmer, J.; Meier, E.A.; Verburg, K.; Skocaj, D.M. Prioritizing Crop Management to Increase Nitrogen Use Efficiency in Australian Sugarcane Crops. *Front. Plant Sci.* **2017**, *8*, 1504. [CrossRef]
5. Liang, C.; Song, J.; Yue, F.; Zhang, J.J. Water environmental protection strategy of Dagu River Basin in Qingdao, China. In *Manufacture Engineering and Environment Engineering*; WIT Press: Qingdao, China, 2014; Volume 1–2, pp. 753–757.
6. Gong, L. Study on Water Pollution Prevention and Water Ecology Improvement Strategies in Dezhou City. Ph.D. Thesis, Ocean University of China, Qingdao, China, 2015. (In Chinese).
7. Xie, L.G. Study on Water Resources Protection and Utilization in the Core Area of Three Gorges Reservoir. Ph.D. Thesis, Chongqing University, Chongqing, China, 2011. (In Chinese).
8. Zhang, J.J. Study on Water Resources Allocation in Small and Medium-Sized Watersheds. Master's Thesis, Chinese Institute of Water Resources and Hydropower Research, Beijing, China, 2018. (In Chinese).
9. Chen, Y.C.; Li, Y.; Yu, H.M.; Qiao, B.; Sun, J.M. Runoff pollutant output characteristics and correlation analysis of typical rainfall paddy fields in Taihu Lake region. *Chin. J. Environ. Eng.* **2016**, *10*, 137–144. (In Chinese)
10. Si, Q.L.; Gao, Y.; Peng, T.; Yu, J.L.; Zhao, W.F. Current status of chemical fertilizer use and progress of research on efficient fertilization for wheat yield increase in China. *Agric. Technol.* **2015**, *35*, 30–31.
11. Wang, T.L.; Cheng, H. Research progress on the current status of nitrogen fertilizer application in agricultural production and its environmental effects. *Heilongjiang Environ. Bull.* **2020**, *33*, 58–59.
12. Zhang, J.; Yin, F.H.; Li, G.Y. Research progress and development trend on the effectiveness of phosphorus fertilizer on farmland. *Sci. Technol. Xinjiang Agric. Reclam.* **2015**, *38*, 57–60.
13. Qu, J.F. Study on the effect of fertilizer application and soil environmental safety. *Phosphate Fertil. Compd. Fertil.* **2010**, *25*, 10–12. (In Chinese)
14. Thompson, R.B.; Martínez-Gaitan, C.; Gallardo, M.; Giménez, C.; Fernández, M.D. Identification of irrigation and N management practices that contribute to nitrate leaching loss from an intensive vegetable production system by use of a comprehensive survey. *Agric. Water Manag.* **2007**, *89*, 261–274. [CrossRef]
15. Thorup-Kristensen, K.; Dorte, B.D.; Hanne, L.K. Crop yield, root growth, and nutrient dynamics in a conventional and three organic cropping systems with different levels of external inputs and N re-cycling through fertility building crops. *Eur. J. Agron.* **2012**, *37*, 66–82. [CrossRef]
16. Stefanelli, D.; Ian, G.; Rod, J. Minimal nitrogen and water use in horticulture: Effects on quality and content of selected nutrients. *Food Res. Int.* **2010**, *43*, 1833–1843. [CrossRef]
17. Colla, G.; Hye-Ji, K.; Kyriacou, M.C.; Youssef, R. Nitrate in fruits and vegetables. *Sci. Hortic.* **2018**, *237*, 221–238. [CrossRef]
18. Hu, Y.C.; Song, Z.W.; Lu, W.L.; Poschenrieder, C.; Schmidhalter, U. Current Soil Nutrient Status of Intensively Managed Greenhouses. *Pedosphere* **2012**, *22*, 825–833. [CrossRef]
19. Li, Y.; Ju, X.T. The key to reduce nitrous oxide emission in farmland is reasonable nitrogen application. *J. Agro-Environ. Sci.* **2020**, *39*, 842–851. (In Chinese)
20. Li, B.; Li, J. Research progress on the prevention and control of agricultural non-point water pollution by microbial technology. *Biotechnol. Bull.* **2015**, *31*, 99–104.
21. Min, J.; Zhang, H.L.; Shi, W.M. Optimizing nitrogen input to reduce nitrate leaching loss in greenhouse vegetable production. *Agric. Water Manag.* **2012**, *111*, 53–59. [CrossRef]
22. Dimzon, I.K.D.; Ann, S.M.; Janine, M.; Roy, K.Y.; Stephan, L.; Heike, W.; Teresita, R.P.; Jutta, M.; Fabian, M.D.; Thomas, P.K. Trace organic chemical pollutants from the lake waters of San Pablo City, Philippines by targeted and non-targeted analysis. *Sci. Total Environ.* **2018**, *639*, 588–595. [CrossRef]
23. Zhang, Z.Y. Analysis of the current situation and progress of agricultural surface source pollution management in water environment. *Green Build. Mater.* **2017**, *4*, 195. (In Chinese)
24. Li, S.M.; Gary, S.B.; Ju, M.; Shi, W.M. Effect of continuous application of inorganic nitrogen fertilizer on selenium concentration in vegetables grown in the Taihu Lake region of China. *Plant Soil* **2015**, *393*, 351–360. [CrossRef]
25. Wang, Y.G.; Wang, H.Y.; Zheng, Y.L.; Sun, X.Y. Advances in research methods and control technologies of agricultural non-point source pollution: A review. *J. China Agric. Resour. Reg. Plan.* **2021**, *42*, 25–33.
26. Li, Q.K.; Hu, Y.W.; Sun, J. Migration and transformation of agricultural non-point source pollutants in drainage ditches. *Chin. J. Eco-Agric.* **2010**, *18*, 210–214. [CrossRef]
27. Yu, Y.L.; Yang, L.Z.; Li, H.N.; Zhu, C.X.; Yang, B.; Xue, L.H. Situation analysis and trend prediction of the prevention and control technologies for planting non-point source pollution. *Environ. Sci.* **2020**, *41*, 3870–3878.
28. Wu, L.; Long, T.Y.; Liu, X. Nitrogen and phosphorus pollution load estimation in small watershed in three gorges reservoir area. *China Water Wastewater* **2012**, *28*, 120–124.
29. Gan, M.Q.; Pod, L.; Huang, Y.; Liu, P.S.; Ye, C.R.; Ma, Y.H. Monitoring and evaluation of surface source pollution from cultivation in Hefei City around Chaohu Lake. *Environ. Monit. Manag. Technol.* **2021**, *33*, 28–32. (In Chinese)
30. Yang, J.L.; Zhang, G.L.; Shi, X.Z.; Wang, H.J.; Cao, Z.H.; Ritsema, C.J. Dynamic changes of nitrogen and phosphorus losses in ephemeral runoff processes by typical storm events in Sichuan Basin, Southwest China. *Soil Tillage Res.* **2009**, *105*, 292–299. [CrossRef]

31. Buchanan, B.P.; Archibald, J.A.; Easton, Z.M.; Shaw, S.B.; Schneider, R.L.; Walter, M.T. A phosphorus index that combines critical source areas and transport pathways using a travel time approach. *J. Hydrol.* **2013**, *486*, 123–135. [CrossRef]
32. Wang, W.G.; Shi, H.B.; Li, X.Y.; Sun, Y.N.; Zhang, W.C.; Zhou, H. Effects of irrigation, fertilization and tillage based on improved Swat model on nutrient load and crop yield in WuliangSuhai Watershed. *Lake Sci.* **2022**, *34*, 1505–1523. (In Chinese)
33. Li, C.; Guo, L.; Ma, Z.W.; Wu, S.; Wu, Z.H.; Ma, Y.H. Research on Reducing fertilizer efficiency and controlling nitrogen and Phosphorus loss in Wuhu City. *Chin. Agric. Sci. Bull.* **2022**, *38*, 100–105. (In Chinese)
34. Tong, R.; Gao, L.; Pang, Z.; Gu, X.; Chen, X.; Zhang, J.; Wang, H.; Wang, S.; Shi, C.; Chen, X. Traceability of Nitrogen and Oxygen Isotopes in Nitrates with the Siar Model: Case Study on the Wangbeng Interval of the Huaihe River Basin. *Pol. J. Environ. Stud.* **2022**, *31*, 2255–2263. [CrossRef]
35. Wang, S.; Wang, Y.; Wang, Y.; Wang, Z. Assessment of Influencing Factors on Non-Point Source Pollution Critical Source Areas in an Agricultural Watershed. *Ecol. Indic.* **2022**, *141*, 109084. [CrossRef]
36. Wang, W.; Li, Z.B.; Shi, P.; Yan, Z.; Pan, B.Z.; Peng, L.; Ding, S.J.; Jing, L.; Bi, Z.L.; Wang, X.K. Vegetation restoration and agricultural management to mitigate nitrogen pollution in the surface waters of the Dan River, China. *Environ. Sci. Pollut. Res.* **2021**, *28*, 47136–47148. [CrossRef]
37. Hou, L.; Zhou, Z.; Wang, R.; Li, J.; Dong, F.; Liu, J. Research on the Non-Point Source Pollution Characteristics of Important Drinking Water Sources. *Water* **2022**, *14*, 211. [CrossRef]
38. Chu, Q.H. Role of fertilization technology in increasing grain yield in China and application prospects. *China Agric. Sci. Technol. Her.* **2013**, *15*, 12–19.
39. Zhang, Q.; Shen, W.; Chu, C.; Shu, L.Z.; Zhou, X.J.; Zhu, S.N. Regulation of nitrogen forms on growth of eggplant under partial root-zone irrigation. *Agric. Water Manag.* **2014**, *142*, 56–65. [CrossRef]
40. Xiao, L.L.; Sun, Q.B.; Yuan, H.T.; Lian, B. A practical soil management to improve soil quality by applying mineral organic fertilizer. *Acta Geochim.* **2017**, *36*, 198–204. [CrossRef]
41. Rodney, B.T.; Luca, I.; Jim, V.R.; Daniele, M. Reducing contamination of water bodies from European vegetable production systems. *Agric. Water Manag.* **2020**, *240*, 106258.
42. Yu, H.B.; Dai, G.J. Migration and transformation mechanism of nitrogen and phosphorus surface source pollution in ditches. *South. Agric.* **2018**, *12*, 103–105. (In Chinese)
43. Nendel, C.; Melzer, D.; Thorburn, P.J. The nitrogen nutrition potential of arable soils. *Sci. Rep.* **2019**, *9*, 5851. [CrossRef]
44. Long, H.Y.; Zhang, D.; Jin, J. Effects of soil texture on growth and development, biomass and soil fertility of Stylosanthes. *Soil* **2017**, *49*, 1049–1052. (In Chinese)
45. Chen, H.L.; Hu, J.W.; Gou, M.M.; Wang, N.; Zhu, S.F.; Xiao, W.F.; Liu, C.F. Soil nutrient characteristics and fertility evaluation of Zigui Citrus Orchard in Hubei Province. *Terr. Ecosyst. Conserv.* **2022**, *2*, 21–31. (In Chinese)
46. Chen, L.H.; Liu, L.Y.; Lv, X.; Li, Y.R. Relationship between rhizosphere soil microbial quantity and fertility of tea tree in different planting years. *Fujian J. Agric. Sci.* **2019**, *34*, 1433–1439. (In Chinese)
47. Liu, L.; Huang, B.J.; Sun, J.; Guo, S.R.; Li, L.Q.; Guo, H.W. Relationship between microbial quantity, enzyme activity and soil fertility in greenhouse pepper continuous cropping soil. *Soil Fertil. Sci. China* **2013**, *2*, 5–10. (In Chinese)
48. Francesco, T.; Stefaan, D.N.; Janjo, D.H.; Hanne, L.K. Nitrogen management of vegetable crops. *Agric. Water Manag.* **2020**, *240*, 106316.
49. Ashraf, M.Y.; Hussain, F.; Akhter, J.; Gul, A.; Ross, M.; Ebert, G. Effect of Different Sources and Rates of Nitrogen and Supra Optimal Level of Potassium Fertilization on Growth, Yield and Nutrient Uptake by Sugarcane. Grown Under Saline Conditions. *Pak. J. Bot.* **2008**, *40*, 1521–1531.
50. Baque, M.A.; Karim, M.A.; Hamid, A.; Hidaka, T. Effects of fertilizer potassium on growth, yield and nutrient uptake of wheat (Triticum aestivum) under water stress conditions. *South Pac. Stud.* **2006**, *27*, 25–35.
51. Noack, S.R.; McBeath, T.M.; McLaughlin, M.J. Potential for foliar phosphorus fertilisation of dryland cereal crops: A review. *Crop. Pasture Sci.* **2010**, *61*, 659–669. [CrossRef]
52. Ashraf, M.Y.; Akhtar, K.; Sarwar, G.; Ashraf, M. Role of the rooting system in salt tolerance potential of different guar accessions. *Agron. Sustain. Dev.* **2005**, *25*, 243–249. [CrossRef]
53. Bastida, F.; Torres, I.F.; Romero-Trigueros, C.; Baldrian, P.; Větrovsky, T.; Bayona, J.M.; Alarcón, J.J.; Hernández, T.; García, C.; Nicolás, E. Combined effects of reduced irrigation and water quality on the soil microbial community of a citrus orchard under semi-arid conditions. *Soil Biol. Biochem.* **2017**, *104*, 226–237. [CrossRef]
54. Yang, L.Z.; Zhou, X.P.; Wang, J.G.; Wang, D.J.; Shi, W.M.; Shi, L.X. Ecological interceptor ditch system and its effectiveness for non-point source pollution control in agricultural fields. *J. Ecol.* **2005**, *11*, 121–124. (In Chinese)

processes

MDPI

Article

Effective Removal of Ammonium from Aqueous Solution by Ball-Milled Biochar Modified with NaOH

Hefeng Yang [1,2], Xiangming Li [1], Yuting Wang [1], Junxia Wang [1], Lihong Yang [2,3], Zhiqiang Ma [2], Jipeng Luo [4,*], Xiaoqiang Cui [1,*], Beibei Yan [1] and Guanyi Chen [1,5]

[1] Tianjin Key Lab of Biomass Waste Utilization, School of Environmental Science and Engineering, Tianjin University, Tianjin 300072, China
[2] CECEP DADI Environmental Remediation Co., Ltd., Beijing 100082, China
[3] School of Environmental Science and Engineering, Southern University of Science and Technology, Shenzhen 518055, China
[4] Key Laboratory of Environmental Remediation and Ecological Health, Ministry of Education, College of Environmental and Resource Sciences, Zhejiang University, Hangzhou 310058, China
[5] School of Mechanical Engineering, Tianjin University of Commerce, Tianjin 300134, China
* Correspondence: luojp@zju.edu.cn (J.L.); cuixiaoqiang@tju.edu.cn (X.C.)

Abstract: The objective of this study was to investigate the feasibility of using modified biochars to enhance removal of ammonium from aqueous solution. The pristine, NaOH-modified, ball-milled, and NaOH-modified ball-milled biochars were prepared from wheat straw at 500 °C. The surface morphology and characteristics of biochar were obviously changed after modification. The NaOH-modification elevated the pH value and ash content of biochar, and the ball-milling treatment promoted the formation of oxygen-containing functional groups. The specific surface area of biochar (20.9 m^2/g) increased to 51.4 m^2/g and 145.6 m^2/g after NaOH-modification and ball-milling treatment, respectively. The modified biochars showed considerable ammonium sorption capacity in a wide pH range (3–7), and the optimal pH of ammonium sorption was around 6. Both NaOH-modification and ball-milling treatment improved ammonium sorption on the biochars. Ammonium sorption of the biochars could be well fitted by the Langmuir and pseudo-second-order model, and the NaOH-modified ball-milled biochar showed the highest ammonium sorption capacity of 8.93 mg g^{-1}. The surface complexation with oxygen-containing functional groups and cation exchange were the dominant mechanisms of ammonium sorption on the biochars. These results indicate that NaOH-modified/ball-milled biochar has a good potential to be used for the ammonium removal from polluted water.

Keywords: ammonium; biochar; sorption; modification; ball milling

1. Introduction

Nitrogen (N) is an essential element for plant growth and human health. In order to achieve a higher crop yield in response to the increasing population, more N-fertilizers have been produced and applied in the soils. However, the chemical N-fertilizer used in soils could release ammonium to water via runoff and leaching, resulting in serious eutrophication of surface water [1–3]. According to the environmental status bulletin of China in 2020, eutrophic surface water (i.e., reservoirs and lakes) accounted for 29.0% of the total water bodies [4]. Hence, several techniques have been developed to remove ammonium from water, such as sorption, ion exchange, biofilm reactors, membrane separation, and phytoremediation [5–9]. Among these removal approaches, sorption has been extensively employed due to its cost-effectiveness and environmental friendliness. For the sorption scheme, the sorbent was crucial to the pollutant removal efficiency, and thus the green sorbent with low cost and high efficiency is urgently needed for ammonium removal.

Biochar is a carbonaceous solid material derived from the pyrolysis of biomass under oxygen-limited conditions [10]. Biochar has shown great potential in the fields of soil improvement, carbon sequestration, and environmental remediation [11–14]. Recently, it was applied to remove ammonium from aqueous solution as a potential sorbent, and the ammonium sorption capacity of biochar was determined by the feedstock properties and production parameters (e.g., pyrolysis temperature, heating rate, and residence time) [15–17]. Cui et al. compared the ammonium sorption capacity of biochars derived from different species of aquatic plants at 500 °C and found that the *Canna indica*-derived biochar had the highest ammonium sorption capacity due to higher content of oxygen-containing functional groups [5]. Ion exchange is also an important mechanism for ammonium adsorption. Gai et al. indicated that the biochar with the highest cation exchange capacity showed the highest ammonium removal efficiency [18]. Tang et al. investigated the effect of pyrolysis temperature (350–550 °C) on the ammonium sorption capacity of biochar and indicated that biochar produced at 450 °C showed the greatest ammonium sorption capacity (1.4 mg/g) owing to its higher functional group density and specific surface area [16]. Notably, the biochars after ammonium sorption could be reused as additive or organic fertilizer for soil improvement [19], providing a potentially cyclic utilization scheme of nutrient. Nevertheless, considering the limited ammonium sorption capacity of raw biomass-derived biochar, it is necessary to modify the biochars to effectively remove more ammonium from water.

In order to improve the ammonium sorption capacity of biochar, several modification strategies have been investigated, including metal oxides/metal doping, alkaline treatment, and granulation [15,20–22]. Vu et al. reported that the corncob derived biochar showed a highest adsorption capacity of 22.6 mg NH_4^+-N/g after modification with HNO_3 and NaOH [22]. Wang et al. found that the $FeCl_3$-HCl modification improved the ammonium sorption capacity of biochar by 14% owing to the increased specific surface area and functional groups (−OH and O−C=O) [23]. Moreover, ball milling has emerged as an effective approach for the syntheses and modification of carbon/metal nanomaterials due to its merits of low cost, high efficiency, and environmental friendliness [24,25]. Recently, ball-milling is considered a promising modification method to enhance the physicochemical properties of biochar via improving the specific surface area and creating new functional groups/defects on the surface [25–27], which are expected to improve the ammonium sorption capacity of biochars. However, quite limited study determined the potential of ball-milled biochar in ammonium removal from water [28]. Moreover, no study investigates the feasibility of combining ball-milling with other modification methods to enhance the ammonium sorption capacity of biochar, and the corresponding ammonium sorption mechanisms of dually modified biochar should be clarified.

In the present study, the dually modified biochars were synthesized and evaluated for their ammonium sorption capacity. The pristine biochar was prepared from wheat straw by pyrolysis at 500 °C, and the modified biochars were treated with alkali (NaOH) and ball-milling. After systematical characterization of the raw and modified biochars, sorption kinetics and isotherm of ammonium on biochars were conducted. The objectives of this study were to (1) develop a feasible modification strategy of biochar for ammonium removal, (2) characterize the physicochemical properties of the modified biochars, and (3) determine the sorption capacity and mechanisms of the modified biochars.

2. Materials and Methods

2.1. Preparation of the Pristine and Modified Biochars

The dried wheat straw samples were ground into ~1 mm powder using a stainless grinding machine for the biochar production. The production of wheat straw biochar was carried out in a modified muffle furnace (Sante Furnace 30–3000 °C, SAF-Therm, Henan, China) under N_2 atmosphere, and the pyrolysis of biomass (10 g) was programmed to drive the temperature to 500 °C at a rate of 5 °C/min and held for 2 h. The solid residues from pyrolysis were ground in a mortar and sieved through a 20-mesh sieve prior to use, and the product was named as wheat straw biochar (WB). For the preparation of alkali modified biochar, the wheat straw was added into 1 mol/L NaOH at a ratio of 1 g:1 mL, and the mixed solution was held for 12 h after mixing well. The alkali modified wheat straw biochar (AWB) was obtained by pyrolyzing the abovementioned alkali modified wheat straw under the same pyrolysis conditions. In order to prepare ball-milled biochars, WB and AWB were placed in the agate jars of a planetary ball mill (DECO-PBM-1L-A, Jiangsu, China). Ball milling was operated at 500 rpm with a mass ratio of 1:20 (biochar to ball) for 8 h, and the direction of rotation changed every 10 min. The derived ball-milled biochars were referred to as ball-milled wheat straw biochar (B-WB) and ball-milled alkali-modified wheat straw biochar (B-AWB), respectively.

2.2. Characterization

Elemental contents (carbon, hydrogen, and nitrogen) of biochar samples were analyzed using a UNICUBE analyzer (Elementar). Fourier transform infrared (FTIR) analysis was employed to determine the functional group composition of biochar in the wavenumber range of 500–4000 cm^{-1} in the form of KBr tablet using an FTIR spectrophotometer (IRAffinity-1S). In order to identify the crystalline structures of biochar, X-ray diffraction analysis (XRD, X' Pert Pro MPD) was carried out at 40 mA and 40 kV using Cu Ka radiation, and the scanning angle was 5–90°. The ash content was measured by placing biochar samples in a muffle furnace and heating to 750 °C for 5 h. The pH of biochar was measured using a pH meter (PHSJ-3F, INESA, Shanghai, China) by adding samples to deionized water at a mass/water ratio of 1:20. The size distribution of biochar was determined using a Malvern Mastersizer.

2.3. Sorption Experiment

Ammonium stock solutions were prepared by dissolving ammonium chloride (NH_4Cl) in deionized water. In order to investigate the effect of pH on the ammonium sorption, the biochar (0.1 g) was added into 30 mL solutions containing 30 mg ammonium-N L^{-1} with an initial pH range between 3.0 and 7.0, which was adjusted by using 0.5 M HCl or NaOH solutions. The final pH in the solutions and the ammonium sorption capacity of biochar were determined after shaken at 120 rpm and 25 °C for 24 h. Sorption kinetics of ammonium was determined in the solutions containing 30 mg ammonium-N L^{-1} under the aforementioned conditions, and the samples were taken and analyzed at various time intervals (0, 5 min, 15 min, 30 min, 1 h, 2 h, 4 h, 7 h, 12 h, and 24 h). Sorption isotherm of ammonium was investigated in the solutions containing different ammonium-N concentrations (0–200 mg/L) under the aforementioned conditions. The biochars after ammonium-N sorption are named as WB-N, AWB-N, B-WB-N, and B-AWB-N, respectively. The biochars after ammonium-N sorption were filtrated and dried at 80 °C for FTIR analysis. The concentrations of ammonium in the solutions were analyzed by colorimetric method. The amount of ammonium sorbed on biochar (Q_t) was calculated by the ammonium-N concentration difference between the initial and equilibrium solutions:

$$Q_t = (C_0 - C_t)V/m \quad (1)$$

where C_0 and C_t (mg/L) are the ammonium-N concentrations at initial and time t, respectively, V is the volume of the aqueous solution (L), and m is the dry weight of biochar (g) used in the sorption experiment. The contribution of cation exchange to the ammonium sorption was investigated in the solution containing 200 mg ammonium-N L^{-1} and deionized water, which was described in our previous study [5].

2.4. Statistical Analysis

Statistical analysis was conducted by the SPSS 20.0, and the mean value and standard deviation were calculated via descriptive statistics. The data were tested at the significant level of $p < 0.05$ using one-way ANOVA. The correlation analysis was conducted with the Pearson test at $p < 0.05$ by SPSS 22.0.

3. Results and Discussion

3.1. Effects of Modification on Biochar Properties

The basic physicochemical characteristics of the raw and modified biochars derived from wheat straw are listed in Table 1. The NaOH-modified biochar (AWB) showed a higher pH (11.0) than the raw wheat straw biochar (WB, pH = 10.3), which resulted from the addition of NaOH during the pyrolysis process. Accordingly, the ash content of AWB (54.22%) was significantly higher than WB (50.19%) ($p < 0.01$). After ball-milling treatment, the pH value of B-WB and B-AWB decreased to 9.3 and 10.5, respectively. The reduction in the pH values of ball-milled biochars could be ascribed to the formation of acidic functional groups (e.g., −COOH) during the ball milling process. In comparison with WB and AWB, B-WB and B-AWB had higher H/C and O/C ratios, indicating that the ball milling favored the formation of less aromatic structures and hydrophilic surfaces of biochar, in agreement with the previous study [25]. The hydrophilic property of ball-milled biochar is expected to promote the sorption of the polar contaminants such as ammonium.

Table 1. The basic physiochemical properties of the raw and modified biochars.

Sample [a]	pH	Ash	N	C (%)	H	O	O/C	Atomic Ratio H/C	(O + N)/C	SSA (m^2/g)
WB	10.25 ± 0.02	50.19 ± 0.37	0.82	38.16	1.764	8.62	0.17	0.55	0.19	20.9
AWB	11.01 ± 0.00	54.22 ± 0.75	0.46	32.18	1.583	11.44	0.27	0.59	0.28	51.4
B-WB	9.25 ± 0.01	48.79 ± 0.20	0.81	37.74	1.978	10.38	0.21	0.63	0.22	145.6
B-AWB	10.53 ± 0.01	51.06 ± 0.26	0.53	33.11	1.884	13.34	0.30	0.68	0.32	121.7

[a]: WB, wheat straw biochar; AWB, NaOH-modified wheat straw biochar; B-WB, ball-milled wheat straw biochar; B-AWB, ball-milled and NaOH-modified wheat straw biochar.

As presented in Table 1, the specific surface area of biochar ranged from 20.9 to 145.6 m^2/g depending on the modification technique. The specific surface area elevated from 20.9 to 51.4 m^2/g after NaOH-modification, and this value increased to 145.6 m^2/g after ball-milling treatment. However, the specific surface area of B-AWB (121.7 m^2/g) is lower than that of B-WB. The decreased specific surface area of B-AWB was probably due to the merging of micropore and a partial occlusion of the pores by the additive during the ball-milling process. This implies that both NaOH-modification and ball milling increased the specific surface area of biochar, which is consistent with the previous studies [1,25,29,30]. Notably, ball milling showed a much better potential to elevate the specific surface area of biochar due to the mechanism difference. The addition of NaOH could create new pores and widen the existing pores [29], while ball milling could increase the external surface area by reducing particle size and elevate the internal surface area via improving internal porosity [25]. The surface morphology of the pristine and modified biochars was presented in the SEM images (Figure 1). In comparison with the rough surface of the unmilled biochars (WB and AWB), the ball-milled biochars had more ultrafine particles, which was consistent with the size distribution results (Figure 2).

Figure 3a illustrates the FTIR spectra of raw and modified biochars. A wide spectrum of the functional group peaks, including −OH at around 3400 cm^{-1}, aromatic C=C/C=O at 1603 cm^{-1}, CH_2 at 1376 and 1436 cm^{-1}, and C−O/C−O−C at 1090 cm^{-1} were well preserved on the surface of the raw biochar. After modification with NaOH and ball-milling treatments, more functional group peaks were observed in the FTIR spectra of AWB, B-WB, and B-AWB. After ball milling of biochar, the corresponding vibration peaks at ~3400, 1603, and 1090 cm^{-1} were significantly enhanced in the spectrum of B-WB (Figure 3a), indicating that ball milling could elevate the contents of oxygen-containing functional groups (e.g., −OH, C=O, and C−O) in the biochar, which is consistent with the element analysis results (Table 1) and the previous studies [1,27]. Similarly, more oxygen-containing functional groups were retained in B-AWB compared with AWB, and the enhancement of oxygen-containing functional groups could promote the sorption capacity of ammonium [5,23]. As depicted in Figure 3b, SiO_2 and KCl were the dominant mineral crystals in the XRD pattern of WB. After NaOH-modification and ball-milling treatments, the intensity of SiO_2 diffraction peak increased in the modified biochars, implying that both NaOH modification and ball milling favored the development of SiO_2 crystal, in good agreement with the FTIR results (Figure 3b).

Figure 1. SEM images of the raw and modified biochars. WB, wheat straw biochar; AWB, NaOH-modified wheat straw biochar; B-WB, ball-milled wheat straw biochar; B-AWB, ball-milled and NaOH-modified wheat straw biochar.

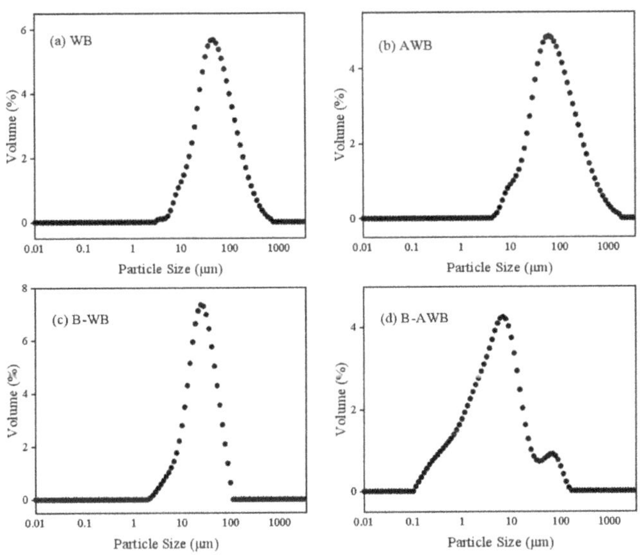

Figure 2. The size distribution of the raw and modified biochars. (**a**) WB, wheat straw biochar; (**b**) AWB, NaOH-modified wheat straw biochar; (**c**) B-WB, ball-milled wheat straw biochar; (**d**) B-AWB, ball-milled and NaOH-modified wheat straw biochar.

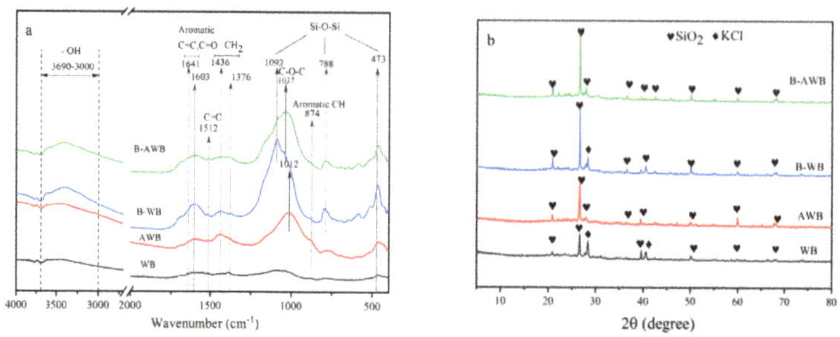

Figure 3. (**a**) The FTIR spectra of the raw and modified biochars; (**b**) XRD patterns of the raw and modified biochars.

3.2. *Ammonium Sorption*

3.2.1. pH Dependent Sorption

The effect of aqueous solution pH on ammonium sorption onto the raw and modified biochars is presented in Figure 4. The results demonstrate that the ammonium sorption capacity of biochars was pH dependent. Generally, the modified biochars showed considerable ammonium sorption capacity in a wide pH range, and the optimal pH of ammonium sorption was around 6 in this study. As depicted in Figure 4, the amount of adsorbed ammonium increased with increasing pH from 3 to 6 and slightly declined as pH elevated to 7. Similar trends were found in the previous studies [15,16]. The lower ammonium sorption capacity at lower pH could be attributed to the competition between NH_4^+ and H^+ for the interaction with the functional groups on the biochar surface. This competition was weakened with the decrease of H^+ at higher pH, and more ammonium was adsorbed onto the biochars. Additionally, the more negatively charged surface of biochar at higher pH of aqueous solution favored the ammonium adsorption via electrostatic attraction [1,22]. The decrease of ammonium removal at pH > 6 could be ascribed to the fact that some of the ammonium is converted to $NH_3{\cdot}H_2O$, which suppressed the sorption process [21]. In comparison with the raw biochar, the ammonium sorption capacity of the modified biochars was more sensitive to the pH change. This phenomenon could be attributed to the more abundant oxygen-containing functional groups on the modified biochars. The interaction between the oxygen-containing functional groups and ammonium was considered a main mechanism of ammonium sorption of biochar [5,15,23], and the chemical state of the oxygen-containing functional groups on the biochar surface was greatly affected by the pH of aqueous solution [31]. For instance, carboxyl groups are acidic groups generally having pKa < 5, and these groups could dissociate with increasing pH.

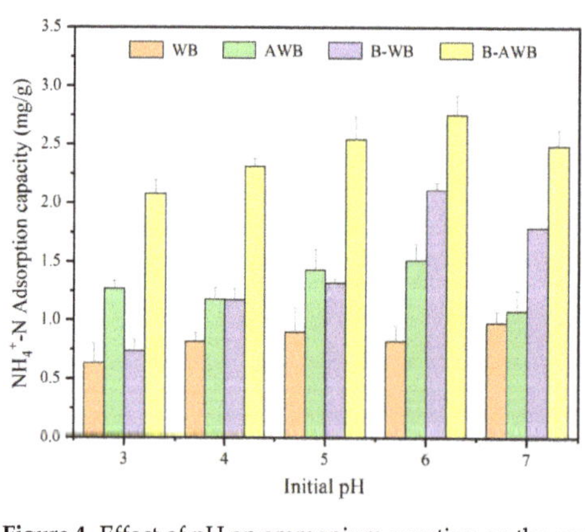

Figure 4. Effect of pH on ammonium sorption on the raw and modified biochars.

3.2.2. Ammonium Sorption Kinetics

Kinetics models were employed to study the reaction dynamics of ammonium sorption onto the raw and modified biochars. As presented in Figure 5, the ammonium sorption onto the raw biochar sharply elevated in the first 1 h, accounting for around 93.2% of the final sorption capacity, and reached saturation in 7 h. Similarly, the ammonium sorption equilibrium was reached within 4 h for the modified biochars. To further analyze the ammonium sorption process onto different biochars, the pseudo-first-order and the pseudo-second-order sorption kinetics models were employed to fit the experimental data. As shown in Table 2, the pseudo-first-order model (R^2 = 0.95) better fitted the ammonium sorption data for the raw biochar than that of the pseudo-second-order model (R^2 = 0.92), while for the NaOH-modified and ball-milled biochars, the fitting degree of the pseudo-second-order model ($0.95 \leq R^2 \leq 0.98$) was better than that of the pseudo-first-order model ($0.94 \leq R^2 \leq 0.98$). Additionally, the predicted ammonium-N sorption capacity of the modified biochars (1.12–2.12 mg/g) by the pseudo-second-order kinetics model was more close to the experimental values (1.14–2.17 mg/g). The fitting results implies that the ammonium sorption onto the modified biochars was mainly dominated by chemical interaction, which is in accordance with the previous studies [5,16], further confirming that more active sites (e.g., oxygen-containing functional groups) for ammonium sorption were successfully introduced after NaOH-modification and ball-milling treatments. Furthermore, the sorption rate constant K_1/K_2 fitted by the sorption kinetics of the modified biochars was larger than that of the raw biochars, in a good agreement with the experimental phenomenon, indicating good application potential of the modified biochars.

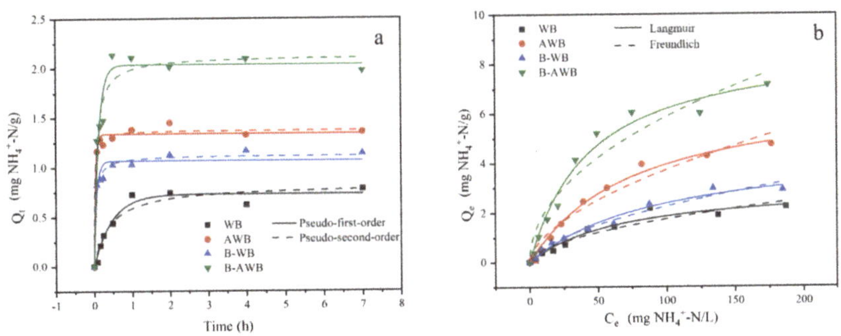

Figure 5. (**a**) Sorption kinetics and of ammonium on the raw and modified biochars; (**b**) isotherm of ammonium on the raw and modified biochars. The symbols and lines are the experimental and modeled results, respectively. (ammonium-N concentration, 0–200 mg/L; adsorbent dose, 3.3 g/L; adsorption time, 0–7 h; adsorption temperature, 25 °C).

Table 2. Sorption parameters of ammonium on the raw and modified biochars obtained from the different kinetics models.

Biochar	Pseudo-First-Order [a]			Pseudo-Second-Order [b]		
	$q_t=q_e(1-e^{-k_1t})$			$q_t=k_2q_e^2t/(1+k_2q_et)$		
	K_1 (h^{-1})	Q_e (mg/g)	R^2	K_2 (g/mg·h)	Q_e (mg/g)	R^2
WB	2.13	0.73	0.95	3.11	0.82	0.92
AWB	23.57	1.34	0.98	46.56	1.37	0.98
B-WB	13.86	1.06	0.94	22.54	1.12	0.98
B-AWB	7.98	2.04	0.94	7.09	2.12	0.95

[a]: Q_t is the amount of ammonium sorption at equilibrium, mg g^{-1}, K_1 is the rate constant of pseudo-first-order model (h^{-1}), Q_e is the sorption capacity at equilibrium calculated by the pseudo-first-order model, mg g^{-1}; [b]: Q_t is the amount of ammonium sorption at equilibrium, mg g^{-1}, K_2 is the rate constant of the pseudo-second-order reaction (g mg^{-1} h^{-1}), Q_e is the sorption capacity at equilibrium calculated by the pseudo-second-order model, mg g^{-1}.

3.2.3. Ammonium Sorption Isotherm

The sorption isotherms of ammonium onto the raw and modified biochars are shown in Figure 5b, and the fitted parameters are summarized in Table 3. The amount of adsorbed ammonium-N by biochar was elevated with increasing initial ammonium concentration. In comparison with the raw biochar, the modified biochars exhibited higher sorption capacity for ammonium-N. The ammonium adsorption behaviors were described using the Langmuir and Freundlich adsorption isotherm models. The Langmuir equation fitted the ammonium sorption data of the raw and modified biochars better than the Freundlich equation, with correlation coefficients (R^2) in the range of 0.95–0.99. The better fitting of sorption data to the Langmuir model implies that the ammonium adsorption occurred largely in a monolayer on the homogeneous surface of biochar, which is consistent with the previous studies [5,16]. The maximum sorption of ammonium-N (Q_m) on B-AWB was 8.93 mg g^{-1}, approximately three times that of WB (3.09 mg g^{-1}). In comparison, B-WB (4.64 mg g^{-1}) and AWB (6.93 mg g^{-1}) had moderate maximum sorption capacity for ammonium-N according to the Langmuir model. These results indicate that both NaOH modification and ball milling improved the ammonium-N sorption capacity of biochar, and the dual modification exhibited enhancement to the ammonium sorption. The maximum ammonium-N sorption capacity of B-AWB is higher than that of biochars derived from rice husk (3.24 mg/g), pine sawdust (3.37–5.38 mg/g), and giant reed (1.21–1.49 mg/g) in the previous studies [32–34], indicating that the modified wheat straw biochar has a good potential for the treatment of ammonium-contaminated water. Furthermore, Table 4 summarizes the ammonium sorption capacity of different biochars in the previous studies.

Table 3. Sorption parameters of ammonium on the raw and modified biochars obtained from the different isotherm models.

Biochar	Langmuir [a] $q_e=K_LC_eq_m/(1+K_LC_e)$			Freundlich [b] $q_e=K_FC_e^N$		
	K_L (L/mg)	Q_m (mg/g)	R^2	K_F (mg/g)	N	R^2
WB	0.02	3.09	0.95	0.15	0.53	0.89
AWB	0.01	6.93	0.99	0.26	0.57	0.95
B-WB	0.03	4.64	0.98	0.16	0.58	0.96
B-AWB	0.02	8.93	0.98	0.65	0.48	0.92

[a]: Q_m is the maximum sorption capacity (mg g^{-1}), C_e is the concentration of amomonium after sorption, K_L is the affinity coefficient of Langmuir model; [b]: Q_m is the maximum sorption capacity (mg g^{-1}), C_e is the concentration of ammonium after sorption, K_F is the experimentally derived capacity coefficient of Freundlich model (mg g^{-1}), N is experimentally derived exponent of Freundlich model.

Table 4. Ammonium sorption capacity of different biochars in previous study.

Adsorbents	Modification Method	Solid-Liquid Ratio (g/L)	Sorption Capacity (mg/g)
Bamboo biochar [15]	Unmodified	1.00	4.96
Digested sludge [16]	Unmodified	10.00	1.40
Rice straw [20]	Unmodified	1.60	3.20
Rice straw [20]	Potassium-iron modified	1.60	10.77
Pine wood chips biochar [35]	Unmodified	1.00	2.86
Reed waste biochar [36]	Unmodified	5.00	1.92
Reed waste biochar [36]	Sulfonation modified	10.00	5.19
Orange peel [37]	Unmodified	10.00	4.36
B-AWB (this study)	NaOH-ball-milled modified	3.33	8.93

3.3. *Ammonium Sorption Mechanisms on the Modified Biochars*

The oxygen-containing functional groups were crucial to the ammonium sorption onto biochars [5,20,23,38]. As depicted in Figure 3a, more oxygen-containing functional groups (e.g., −OH, C=O, and C−O) were formed in the biochar after NaOH modification and ball milling, and the ammonium-N sorption capacity of the modified biochars was greatly enhanced (Figure 5b). Correspondingly, B-AWB with most abundant oxygen-containing functional groups exhibited the highest ammonium-N sorption capacity. Moreover, the amount of ammonium-N sorbed on biochars was significantly correlated with the O/C ratio of biochar ($r = 0.993$, $p < 0.01$), implying that oxygen-containing functional groups is involved in ammonium sorption onto biochars. As presented in Figure 6, obvious changes could be observed on the FTIR spectra of biochar at the peaks of C-O-C and C = O after ammonium-N sorption, which is consistent with the previous studies [5,23]. These results demonstrate that the oxygen-containing functional groups play a dominant role in the ammonium sorption onto biochar.

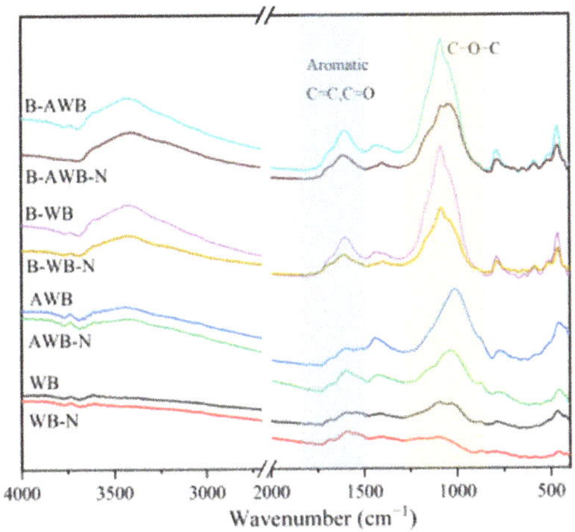

Figure 6. FTIR spectra of biochars before and after ammonium sorption.

In addition to the contribution of oxygen-containing functional groups, the role of cation exchange is also should be taken into consideration for ammonium sorption of biochars. A significant amount of metal ions (e.g., K^+, Na^+, Ca^{2+}, and Mg^{2+}) remained on the biochars via forming complexes with oxygen-containing functional groups, electrostatic attraction, and precipitation. Cui et al. reported that the ammonium-N sorption capacity of biochar was highly correlated with the amount of Mg^{2+} and Ca^{2+} released after ammonium sorption [5], and a similar trend was reported by Zheng et al. [39]. To investigate the role of cation exchange in ammonium sorption, the amount of metal ions released in solution before and after ammonium sorption on biochar was determined in this study. As illustrated in Figure 7, the obvious release of Ca^{2+}, K^+, Mg^{2+}, and Na^+ occurred during the ammonium sorption process. This demonstrates that cation exchange occurred during the ammonium sorption onto the biochars. However, the total amounts of cations released from biochar after sorption were not highly correlated with the ammonium-N sorption capacity of biochar ($p > 0.05$), suggesting that the cation exchange was not the dominant ammonium sorption mechanism of biochars in the present study.

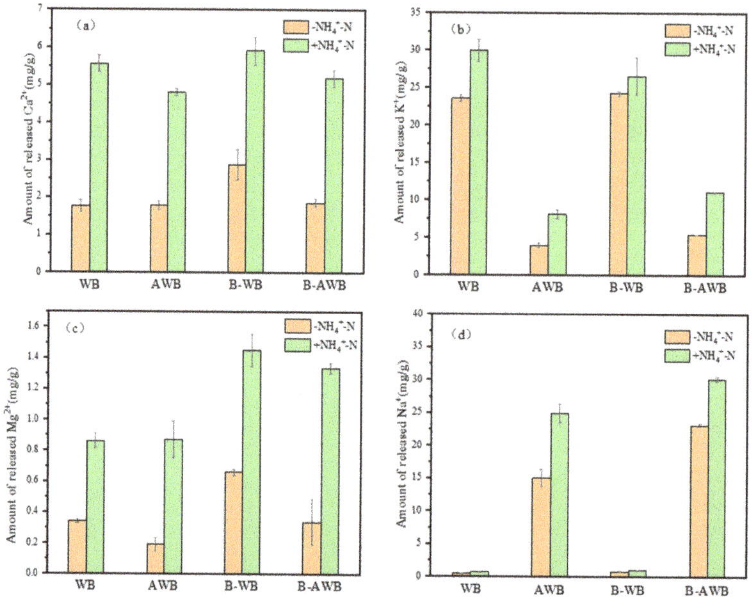

Figure 7. (**a**) The amount of Ca^{2+} released from biochars into solution before ($-NH_4^+-N$) and after ($+NH_4^+-N$) ammonium sorption; (**b**) the amount of K^+ released from biochars into solution before ($-NH_4^+-N$) and after ($+NH_4^+-N$) ammonium sorption; (**c**) the amount of Mg^{2+} released from biochars into solution before ($-NH_4^+-N$) and after ($+NH_4^+-N$) ammonium sorption; (**d**) the amount of Na^+ released from biochars into solution before ($-NH_4^+-N$) and after ($+NH_4^+-N$) ammonium sorption.

4. Conclusions

NaOH-modified/ball-milled biochars derived from wheat straw were prepared and employed for the effective ammonium removal from aqueous solutions, and the dual modification exhibited enhancement to the ammonium sorption. In comparison with the raw biochar, the physicochemical properties of the modified biochars were obviously changed. Considering the low cost of NaOH/ball-milling modifications and wheat straw feedstock, the NaOH-modified/ball-milled wheat straw biochar could be used as a effective adsorbent for ammonium removal. Furthermore, ammonium-N is an essential nutrient for plant growth, and thus the N-laden biochars after ammonium sorption have a good potential as an alternative fertilizer. However, the toxicity assessment and field trials should be conducted before the large-scale soil application of the N-laden biochars after ammonium sorption.

Author Contributions: H.Y.: Conceptualization, Writing—Original draft preparation, Funding acquisition. X.L.: Conceptualization, Methodology, Validation, Formal analysis, Writing—Original draft preparation. Y.W.: Resources, Investigation. J.W.: Software, Visualization. L.Y.: Resources, Data supply. Z.M.: Resources, Data supply. J.L.: Writing—review & editing, Supervision. X.C.: Conceptualization, Methodology, Resources, Writing—original draft & editing, Supervision. B.Y.: Project administration. G.C.: Writing—review & editing, Funding acquisition, Supervision. All authors have read and agreed to the published version of the manuscript.

Funding: This research was funded by Ministry of Science and Technology of the People's Republic of China (2021YFE0102500) and CECEP DADI Environmental Remediation Co., Ltd. (2021GKF-0474).

Data Availability Statement: Data will be made available on request.

Conflicts of Interest: There are no conflict of interest to declare.

References

1. Qin, Y.; Zhu, X.; Su, Q.; Anumah, A.; Gao, B.; Lyu, W.; Zhou, X.; Xing, Y.; Wang, B. Enhanced removal of ammonium from water by ball-milled biochar. *Environ. Geochem. Health* **2020**, *42*, 1579–1587. [CrossRef] [PubMed]
2. Conley, D.J.; Paerl, H.W.; Howarth, R.W.; Boesch, D.F.; Seitzinger, S.P.; Havens, K.E.; Lancelot, C.; Likens, G.E. Controlling Eutrophication: Nitrogen and Phosphorus. *Science* **2009**, *323*, 1014–1015. [CrossRef] [PubMed]
3. Cameron, K.C.; Di, H.J.; Moir, J.L. Nitrogen losses from the soil/plant system: A review. *Ann. Appl. Biol.* **2013**, *162*, 145–173. [CrossRef]
4. MEPC. *China Environmental Status Bulletin in 2020*; Ministry of Environment Protection of the People's Republic of China: Beijing, China, 2021.
5. Cui, X.; Hao, H.; Zhang, C.; He, Z.; Yang, X. Capacity and mechanisms of ammonium and cadmium sorption on different wetland-plant derived biochars. *Sci. Total Environ.* **2016**, *539*, 566–575. [CrossRef] [PubMed]
6. Ji, Z.-Y.; Yuan, J.-S.; Li, X.-G. Removal of ammonium from wastewater using calcium form clinoptilolite. *J. Hazard. Mater.* **2007**, *141*, 483–488. [CrossRef]
7. Chen, S.-T.; Wickramasinghe, S.R.; Qian, X. High Performance Mixed-Matrix Electrospun Membranes for Ammonium Removal from Wastewaters. *Membranes* **2021**, *11*, 440. [CrossRef]
8. Motlagh, A.R.A.; LaPara, T.M.; Semmens, M.J. Ammonium removal in advective-flow membrane-aerated biofilm reactors (AF-MABRs). *J. Membr. Sci.* **2008**, *319*, 76–81.
9. Al-Ajalin, F.A.H.; Abdullah, S.R.S.; Idris, M.; Kurniawan, S.B.; Ramli, N.N.; Imron, M.F. Removal of ammonium, phosphate, and COD by bacteria isolated from *Lepironia articulata* and *Scirpus grossus* root system. *Int. J. Environ. Sci. Technol.* **2022**, *19*, 11893–11904. [CrossRef]
10. IBI. *Standardized Product Definition and Product Testing Guidelines for Biochar That Is Used in Soil*; International Biochar Initiative: Canandaigua, NY, USA, 2012.
11. Ahmad, M.; Rajapaksha, A.U.; Lim, J.E.; Zhang, M.; Bolan, N.; Mohan, D.; Vithanage, M.; Lee, S.S.; Ok, Y.S. Biochar as a sorbent for contaminant management in soil and water: A review. *Chemosphere* **2014**, *99*, 19–33. [CrossRef]
12. Cui, X.; Wang, J.; Wang, X.; Khan, M.B.; Lu, M.; Khan, K.Y.; Song, Y.; He, Z.; Yang, X.; Yan, B.; et al. Biochar from constructed wetland biomass waste: A review of its potential and challenges. *Chemosphere* **2022**, *287*, 132259. [CrossRef]
13. Cui, X.; Yang, X.; Sheng, K.; He, Z.; Chen, G. Transformation of Phosphorus in Wetland Biomass during Pyrolysis and Hydrothermal Treatment. *ACS Sustain. Chem. Eng.* **2019**, *7*, 16520–16528. [CrossRef]
14. Gao, S.; DeLuca, T.H.; Cleveland, C.C. Biochar additions alter phosphorus and nitrogen availability in agricultural ecosystems: A meta-analysis. *Sci. Total Environ.* **2019**, *654*, 463–472. [CrossRef] [PubMed]
15. Fan, R.; Chen, C.-L.; Lin, J.-Y.; Tzeng, J.-H.; Huang, C.-P.; Dong, C. Adsorption characteristics of ammonium ion onto hydrous biochars in dilute aqueous solutions. *Bioresour. Technol.* **2019**, *272*, 465–472. [CrossRef] [PubMed]
16. Tang, Y.; Alam, S.; Konhauser, K.O.; Alessi, D.S.; Xu, S.; Tian, W.; Liu, Y. Influence of pyrolysis temperature on production of digested sludge biochar and its application for ammonium removal from municipal wastewater. *J. Clean. Prod.* **2019**, *209*, 927–936. [CrossRef]
17. Zhao, Y.; Huang, L.; Chen, Y. Biochars derived from giant reed (*Arundo donax* L.) with different treatment: Characterization and ammonium adsorption potential. *Environ. Sci. Pollut. Res.* **2017**, *24*, 25889–25898. [CrossRef]
18. Gai, X.; Wang, H.; Liu, J.; Zhai, L.; Liu, S.; Ren, T.; Liu, H. Effects of feedstock and pyrolysis temperature on biochar adsorption of ammonium and nitrate. *PLoS ONE* **2014**, *9*, e113888. [CrossRef]
19. Bai, X.; Li, Z.; Zhang, Y.; Ni, J.; Wang, X.; Zhou, X. Recovery of Ammonium in Urine by Biochar Derived from Faecal Sludge and its Application as Soil Conditioner. *Waste Biomass Valorization* **2017**, *9*, 1619–1628. [CrossRef]
20. Chandra, S.; Medha, I.; Bhattacharya, J. Potassium-iron rice straw biochar composite for sorption of nitrate, phosphate, and ammonium ions in soil for timely and controlled release. *Sci. Total Environ.* **2020**, *712*, 136337. [CrossRef]
21. Liu, Z.; Xue, Y.; Gao, F.; Cheng, X.; Yang, K. Removal of ammonium from aqueous solutions using alkali-modified biochars. *Chem. Speciat. Bioavailab.* **2016**, *28*, 26–32. [CrossRef]
22. Vu, T.M.; Trinh, V.T.; Doan, D.P.; Van, H.T.; Nguyen, T.V.; Vigneswaran, S.; Ngo, H.H. Removing ammonium from water using modified corncob-biochar. *Sci. Total Environ.* **2017**, *579*, 612–619. [CrossRef]
23. Wang, S.; Ai, S.; Nzediegwu, C.; Kwak, J.-H.; Islam, S.; Li, Y.; Chang, S.X. Carboxyl and hydroxyl groups enhance ammonium adsorption capacity of iron (III) chloride and hydrochloric acid modified biochars. *Bioresour. Technol.* **2020**, *309*, 123390. [CrossRef] [PubMed]
24. Amusat, S.O.; Kebede, T.G.; Dube, S.; Nindi, M.M. Ball-milling synthesis of biochar and biochar–based nanocomposites and prospects for removal of emerging contaminants: A review. *J. Water Process. Eng.* **2021**, *41*, 101993. [CrossRef]
25. Zhang, Q.; Wang, J.; Lyu, H.; Zhao, Q.; Jiang, L.; Liu, L. Ball-milled biochar for galaxolide removal: Sorption performance and governing mechanisms. *Sci. Total Environ.* **2019**, *659*, 1537–1545. [CrossRef] [PubMed]
26. Zheng, Y.; Wan, Y.; Chen, J.; Chen, H.; Gao, B. MgO modified biochar produced through ball milling: A dual-functional adsorbent for removal of different contaminants. *Chemosphere* **2020**, *243*, 125344. [CrossRef]
27. Zhuang, Z.; Wang, L.; Tang, J. Efficient removal of volatile organic compound by ball-milled biochars from different preparing conditions. *J. Hazard. Mater.* **2021**, *406*, 124676. [CrossRef] [PubMed]

28. Feng, Q.; Chen, M.; Wu, P.; Zhang, X.; Wang, S.; Yu, Z.; Wang, B. Simultaneous reclaiming phosphate and ammonium from aqueous solutions by calcium alginate-biochar composite: Sorption performance and governing mechanisms. *Chem. Eng. J.* **2021**, *429*, 132166. [CrossRef]
29. Li, B.; Yang, L.; Wang, C.-Q.; Zhang, Q.-P.; Liu, Q.-C.; Li, Y.-D.; Xiao, R. Adsorption of Cd(II) from aqueous solutions by rape straw biochar derived from different modification processes. *Chemosphere* **2017**, *175*, 332–340. [CrossRef]
30. Bashir, S.; Zhu, J.; Fu, Q.; Hu, H. Comparing the adsorption mechanism of Cd by rice straw pristine and KOH-modified biochar. *Environ. Sci. Pollut. Res.* **2018**, *25*, 11875–11883. [CrossRef]
31. Chen, Z.; Xiao, X.; Chen, B.; Zhu, L. Quantification of chemical states, dissociation constants and contents of oxygen-containing groups on the surface of biochars produced at different temperatures. *Environ. Sci. Technol.* **2015**, *49*, 309–317. [CrossRef]
32. Zhu, K.; Fu, H.; Zhang, J.; Lv, X.; Tang, J.; Xu, X. Studies on removal of NH_4+-N from aqueous solution by using the activated carbons derived from rice husk. *Biomass Bioenergy* **2012**, *43*, 18–25. [CrossRef]
33. Hou, J.; Huang, L.; Yang, Z.; Zhao, Y.; Deng, C.; Chen, Y.; Li, X. Adsorption of ammonium on biochar prepared from giant reed. *Environ. Sci. Pollut. Res.* **2016**, *23*, 19107–19115. [CrossRef] [PubMed]
34. Yang, H.I.; Lou, K.; Rajapaksha, A.U.; Ok, Y.S.; Anyia, A.O.; Chang, S.X. Adsorption of ammonium in aqueous solutions by pine sawdust and wheat straw biochars. *Environ. Sci. Pollut. Res.* **2018**, *25*, 25638–25647. [CrossRef] [PubMed]
35. Rezaee, M.; Gitipour, S.; Sarrafzadeh, M.H. EVALUATION OF PHOSPHATE AND AMMONIUM ADSORPTION DESORPTION OF SLOW PYROLYZED WOOD BIOCHAR. *Environ. Eng. Manag. J.* **2021**, *20*, 217–227. [CrossRef]
36. Zhang, M.; Sun, R.; Song, G.; Wu, L.; Ye, H.; Xu, L.; Parikh, S.J.; Nguyen, T.; Khan, E.; Vithanage, M.; et al. Enhanced removal of ammonium from water using sulfonated reed waste biochar-A lab-scale investigation. *Environ. Pollut.* **2022**, *292*, 118412. [CrossRef]
37. Hu, X.; Zhang, X.; Ngo, H.H.; Guo, W.; Wen, H.; Li, C.; Zhang, Y.; Ma, C. Comparison study on the ammonium adsorption of the biochars derived from different kinds of fruit peel. *Sci. Total Environ.* **2020**, *707*, 135544. [CrossRef]
38. Wang, Z.; Guo, H.; Shen, F.; Yang, G.; Zhang, Y.; Zeng, Y.; Wang, L.; Xiao, H.; Deng, S. Biochar produced from oak sawdust by Lanthanum (La)-involved pyrolysis for adsorption of ammonium (NH_4^+), nitrate (NO_3^-), and phosphate ($PO4^{3-}$). *Chemosphere* **2015**, *119*, 646–653. [CrossRef]
39. Zheng, H.; Wang, Z.; Deng, X.; Zhao, J.; Luo, Y.; Novak, J.; Herbert, S.; Xing, B. Characteristics and nutrient values of biochars produced from giant reed at different temperatures. *Bioresour. Technol.* **2013**, *130*, 463–471. [CrossRef]

processes

Article

Application of Landfill Gas-Water Joint Regulation Technology in Tianjin Landfill

Jun Liu [1], Tianqi Pan [2], Huihui Zhao [2], Yan Guo [2], Guanyi Chen [3] and Li'an Hou [1,*]

1 School of Environmental Science and Engineering, Tianjin University, Tianjin 300072, China; lzriliu@163.com
2 Nanjing Wondux Environmental Protection Technology Co., Ltd., Nanjing 211100, China; pantq@njwds.com (T.P.); zhaohh@njwds.com (H.Z.); guoy@njwds.com (Y.G.)
3 School of Mechanical Engineering, Tianjin University of Commerce, Tianjin 300134, China; chen@tju.edu.cn
* Correspondence: 13912969682@163.com

Abstract: Landfills have long been widely used to dispose of Municipal Solid Waste (MSW). However, many landfills have faced early closure issues in recent years due to overload operations. Although in-situ aeration technology can quickly stabilize MSW, low oxygen utilization rates present a general problem that results in high energy-consuming and operating costs. This research aims to improve oxygen utilization efficiency by observing the dynamic respiratory index and the removal of contaminants. Three continuous reactors were constructed and designed with targeted aeration and re-circulation schemes for different landfill ages. The results show that a well-designed aerobic, semi-aerobic, and anaerobic reactor can fully degrade the organic components of MSW with different landfill ages, and the quantity of waste has been reduced by more than 60%. Additionally, it was disclosed that gas-water joint technology has a promotional effect on activating microorganisms.

Keywords: dynamic respiratory index; aerobic remediation; rapid stabilization; biochemical treatment

1. Introduction

In recent years, the amount of Municipal Solid Waste (MSW) has rapidly increased in both urban and rural areas due to population expansion and urbanization. For instance, over 2000 landfills in China are currently experiencing overload operation and early closure. Because the utilization of aerobic stabilization technology can help to digest organic components in landfill wastes within a short time, this technology has gained increasing attention as a promising method for MSW treatment.

Compared to traditional landfill methods, aerobic stabilization technology offers the advantage of shortening the treatment time, reducing the volume of waste, and minimizing environmental pollution. Aeration can significantly reduce the production of gas and leachate in landfill waste, which is also beneficial for stabilizing the MSW. Read et al. found that aeration and leachate recirculation can rapidly stabilize the waste load, accelerate the settlement of the landfill, reduce methane and leachate production, and significantly reduce the pollutants in the leachate [1]. Radoslaw Slezak et al. studied the degradation process of solid waste in aerobic and anaerobic conditions using a bioreactor at a landfill site [2]. Their research showed that even a small aeration rate is beneficial for the degradation of organic matter in the leachate. The amount of CO_2 and CH_4 released under aerobic conditions amounts to one-fifth of the gas released under anaerobic conditions [2].

The aerobic restoration technology of landfills can be divided into three types: (1) low-pressure continuous aeration and air-lift technology [3]; (2) short-range high-pressure pulse jetting and low-pressure air-lift technology [4]; and (3) continuous low-pressure aeration with passive low-pressure aeration and air-lift [5]. Low-pressure continuous aeration and air-lift technology continuously inject air into the landfill through a vertical well system, transforming the anaerobic environment into an aerobic one. This process supplements the amount of moisture required to facilitate the degradation of organic matter in the landfill. As for the low-pressure aeration/air-lift technology, the gas injected/extracted from the landfill is within a positive/negative pressure of 30 kPa, and the pressure provided is usually within 3~8 kPa [6]. High-pressure aeration technology intermittently injects compressed air into the landfill through specified nozzles and electromagnetic valves. The positive air pressure injected into the landfill can reach a maximum of 600 kPa [4], providing higher aeration efficiency than the low-pressure systems but with higher energy consumption. Passive low-pressure and air-lift systems inject air into the landfill through wind-energy aerodynamic systems, extracting gas from the landfill via wind-energy air-lift devices. Because this system provides limited air volume and pressure but requires less electricity during operation, it greatly reduces operating costs [7]. Several authors have researched the changes in methane gas concentration in different monitoring wells under different negative pressure and flow conditions. Zhou and co-workers found that the radius of influence and the impact on methane concentration varied under varying extraction flow rates. They suggested that the optimal radius of influence for extraction should be between 20 and 25 m, and the optimal negative pressure should be 20 kPa [8]. In addition, deep waste has a higher permeability than shallow waste, and the amount of extraction in the deep layers is greater than that in the shallow layers, resulting in negative pressure inside the waste pile. If the landfill is not properly sealed, the oxygen concentration inside the monitoring wells will be similar to that of the outside, affecting the negative pressure environment inside the waste pile.

Researchers experimentally performed the injection and extraction of air in a landfill in Guiyang; they analyzed the impact of airflow, pressure, and influence of radius on the aerobic stabilization of the waste pile by measuring the methane content and changes in the pressure in the monitoring wells. The study found that as the vacuum in the waste pile increased, the radius of extraction also increased; when the negative pressure was 10 kPa, an effective radius of 20 m was observed. The higher the extraction pressure, the higher the equipment and operating costs. By observing a change in oxygen concentration in the monitoring wells, it was found that the higher the injection pressure, the greater the injection radius; hence, the injection flow rate and pressure were proportional within a certain range. Nevertheless, the study suggested that the best pressure loss for injection wells in the aerobic stabilization process is 30 kPa [9].

The Kuhstedt Landfill in Germany is one of the most well-known landfills in the world that has been treated with aerobic restoration technology. After the restoration system was put into operation, the organic matter in the landfill tended to be stable but could still produce methane [5]. Therefore, in the later stage of the treatment process, a passive low-pressure and gas extraction system (Wi DAVA-System) was used for the long-term passive aeration treatment. The system consists of 12 wind-powered gas extraction devices and two wind-powered gas injection devices, each of which can inject 3–10 m^3 of air into the landfill on average every day. Research data showed that after the system had been running for 12 months, the landfill tended to be stable, and the biodegradable carbon conversion rate was about 90%. Compared with anaerobic degradation, this system can accelerate the degradation rate of organic matter by four to six times. In addition, the Konstanz Dorfweiher Landfill in Germany used continuous low-pressure aeration and short-range high-pressure pulse aeration to treat the landfill. By using high-pressure pulses, the system changed the flow path of air and leachate in the landfill, reducing short-circuiting inside the landfill. The study showed that after low-pressure aeration was carried out in the landfill, the application of high-pressure pulse technology further degraded the organic matter in the landfill [4]. Similarly, in optimizing the aeration method, a landfill site in Florida adopts different depths of aeration wells to aeration for different depths of waste and achieves good aeration effects [10].

However, there are also some challenges associated with aerobic stabilization technology, such as the need for adequate oxygen supply, effective control of the temperature and moisture content, and management of potential odor and leachate problems. Therefore, improving the technology and optimizing the operational parameters are essential to ensure its successful application in MSW treatment. Traditional aerobic bioreactors have used Biological Degradable Material (BDM) or landfill porosity as the aeration rate design index [8]. However, the oxygen utilization coefficient is limited in practical applications due to factors such as varying oxygen demands of different landfill age waste, low aeration efficiency at depths greater than 8 m, the aerated zone, and the short flow phenomenon. Research has shown that the oxygen utilization coefficient in Chinese landfills is 8% to 15%, whereas foreign cases range from 15% to 22% [9]. Therefore, it is necessary to advance the aeration system and improve the efficiency of oxygen utilization depending on different landfills.

According to national regulations, the closure period, organic contents, landfill gas, odor index, and sedimentation rate can be used to determine the stabilization of landfills [5]. Although the MSW has been pretreated by crushing, screening, ramming, etc., the landfill wastes are highly heterogeneous, which will cause sampling fluctuation. The methods used to determine the stabilization of the landfill were considered by Kelly, who found that there is a significant correlation between the ratio of cellulose to lignin and the landfill stabilization process [11]. Similarly, Li proposed that the BDM and the ratio of cellulose to lignin (C/L) can accurately reflect the stabilization process, reducing the interference by waste heterogeneity [12]. Meanwhile, the Respiration Index (RI) has been considered an important index to evaluate the biodegradability of solid waste, which can react with the oxygen consumed by the waste during a specified period.

Considering the differences in the degradation rates of waste from different regions and depths, this paper used comprehensive measures to regulate the aeration mode and volume to improve the anaerobic remediation efficiency of the landfill and introduced the DRI parameter to forecast the changes in the oxygen demand of waste at different landfill depths and landfill ages.

2. Materials and Methods

2.1. Description of the Case Study

The Huaming Landfill is located in Tianjin province. The waste volume buried within the soil is estimated to be within ~600,000 to 700,000 cubic meters. For historical reasons, the local district lacks suitable facilities for the harmless disposal of household waste. The household waste in the local district is mainly disposed of at the Dahanzhuang Landfill, Shuanggang Incineration Plant, and Beichen Shuangkou Landfill. Owing to the rapid development of the city since 2013, the generation of household waste has continuously increased; hence, the Huaming Landfill has become a place where residents living on Huaming Street discard their waste. Moreover, residents from nearby areas also add their own waste. In addition, the Huaming Landfill is a non-standard landfill with no environmental protection facilities such as anti-seepage, leachate treatment, or landfill gas discharge, all of which pose a great hazard to the local atmosphere, the surrounding water environment, and the residents' safety and well-being. Therefore, this project has decided to use an aerobic restoration method to rapidly degrade organic pollutants in the waste to eliminate the negative impact of waste on the surrounding environment in the shortest possible time. The Aerial view of the Huaming landfill site was shown in Figure 1.

This experiment was conducted at a landfill in Tianjin. The landfill covers an area of about 160,000 m^2, with a landfill depth of 11 m, 700,000 m^3 MSW, and 800,000 m^3 leachate. The site has been used for the treatment of MSW since the year 2013.

The experiment was sampled at various depths of 2 m, 5 m, and 9 m at 20 points in different areas of the landfill. The sand-filling method was used to measure the average density of samples. The following table shows the average density of the landfill:

Table 1 shows the trend of average density variation in the landfill increasing along with depth. This increase occurs because the waste in the landfill areas is mainly composed of MSW, which has a high organic content. The decay and the self-weight of waste increase the density of the landfill. The density of waste at a depth of 9 m can reach up to 1024 kg/m^3.

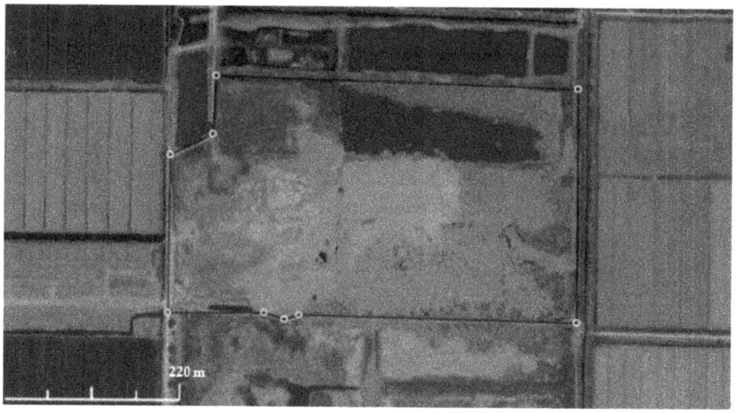

Figure 1. Aerial view of the Huaming landfill site.

Table 1. Analysis of the average density of waste samples at different depths in the landfill.

Depth (m)	2	5	9
Landfill age (a)	1~2	3~5	6~8
Average Density (kg/m^3)	689	812	926

2.2. Bio-Reactor Description and Monitoring Plan

The experiment was designed as a parallel study to investigate the effect of different aeration methods on the rapid stabilization of the Huaming landfill site. It considers the energy consumption and remediation effects in order to determine the best disposal plan for the landfill. The dominant kind of waste in the Huaming landfill is mainly from domestic sources, based on the different depths of the landfill, which is mainly divided into three parts: within two years, two to five years, and five to nine years. For waste sampled at different depths and different landfill ages, the experiment used intermittent aeration to achieve rapid stabilization of the landfill. Fresh waste obtained within two years has a high content of self-generated organic matter and a high BDM, and this was continuously and rapidly aerated to degrade the organic matter present in the waste. Some of the easily degradable organic matter present in the waste obtained within a two to five-year period has been degraded. Limited aeration can create aerobic-anaerobic zones within the landfill, thus increasing the variety of bacteria responsible for degradation to achieve the synergistic removal of carbon and nitrogen. However, at a longer landfill age of five to nine years, most of the organic matter occurring in the waste is difficult to degrade. These types of waste include rubber, plastic, and lignin, which have low degradability. Moreover, according to the results obtained from small-scale experiments, the optimal moisture content for aerobic restoration of waste is in the range of 40–50%, and the optimal oxygen concentration is above 15%. Furthermore, oxygen concentration has a greater effect on waste degradation than moisture content. Therefore, the experiment simulates the aerobic restoration of waste in the landfill for three different landfill ages by setting up three reactors (referred to as M_1, M_2, and M_3) to provide a reference for subsequent research studies.

Establishment of Experiment System

Three bioreactors were set up to evaluate the effect of aerobic stabilization technology on waste bodies and leachate composition at different depths. The effective diameter of the experimental column (D) is 1500 mm, the effective length (h) is 1630 mm, the effective volume of the experimental column (V) is 2.88 m^3, and the average density of the waste is 0.85 t/m^3.

In the experiment, stratified aeration technology was used to stabilize the landfill waste based on the different landfill depths and landfill ages. Considering that the high content of organic matter and the BDM of waste is less than two years of landfill age, continuous aeration will be used to rapidly degrade the organic substances. Even though there is a limited supply of oxygen for waste obtained within two to five years, the aerobic and semi-aerobic areas can be built inside the bioreactor to increase the type of degrading bacteria, which will remove the carbon and nitrogen concurrently. As for long-term landfill waste with a period of five to nine years, the organic substances are rubber, plastic, and lignin; these have low degradability. Consequently, three bioreactors (recorded as M_1, M_2, and M_3) are set in this experiment to simulate the stabilization of waste with different ages in the landfill site and serve as a reference for this project.

The sampling method used in the experiment involves removing large debris (such as stones larger than 50 cm) from the sample, then crushing the remaining waste into small particles ranging from 100 mm to 200 mm using a crusher. Using a layered landfill and manual compaction method, each landfill of 30 cm was compacted.

For the aeration, the reactor used a blower to aerate the waste in the reactor through a buried aeration pipe 50 cm long and perforated 1 m away from the reactor surface. The rector (M_1) is filled with landfill waste with less than two years of landfill age that is sufficiently aerated with air, whereas reactor (M_2) is filled with landfill waste of two to three years of landfill age intermittently aerated with air in the reactor every two days. Meanwhile, reactor (M_3) is filled with landfill waste of five to nine years of landfill age without an aeration process.

A spray nozzle is installed at the top of the reactor, and a plurality of fan-shaped nozzles is used. A peristaltic pump was used to spray the leachate from the top of the reactor. The number and angle of the nozzles are adjusted to achieve the design area of the spray. At the same time, leachate obtained from the landfill is used to backfill the M_1, M_2, and M_3 reactors. As the M_1 continues to produce leachate, the leachate generated in the reactor is backfilled to M_2, whereas the leachate produced from the M_2 reactor is backfilled to the M_3 reactor based on the design value. If the leachate produced is not sufficient to meet the design value, the leachate obtained from the landfill site is used to supplement the remaining part. The relevant system is shown in Figures 2 and 3 below:

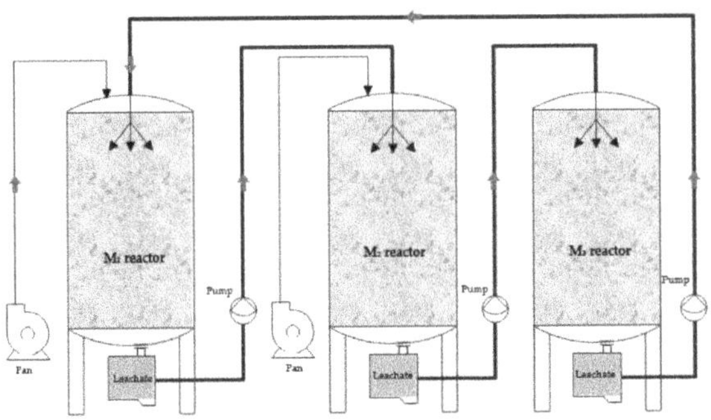

Figure 2. The process of experimental system.

The experiment took samples of waste from the landfill at different depths of 0–2 m, 4–5 m, and 5–9 m. After the removal of large stones (>30 cm in diameter) and other impurities, the samples were filled into the reactors in layers and compacted at every 20 cm of depth. When the reactors were filled to the monitoring probe and vent pipe areas, the probes and pipes were buried in place, and the waste was compacted into the reactors. Because the waste at 5–9 m is older and more decomposed, the density of the M_3 reactor was found to be 1.1 t/m^3, while the densities of the M_1 and M_2 reactors were 0.85 t/m^3, as shown in Table 2.

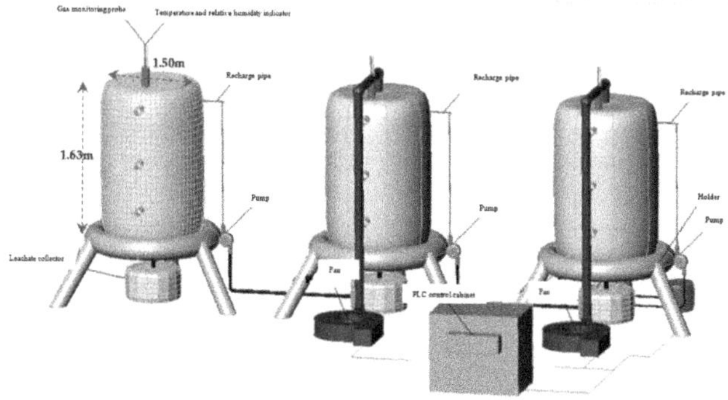

Figure 3. The main view of experimental system.

Table 2. Reactor waste loading parameters.

Number	Landfill Age (a)	Mass (t)	Density (t/m^3)
M_1	0~2	2.45 ± 0.01	0.85 ± 0.1
M_2	2~5	2.45 ± 0.01	0.85 ± 0.1
M_3	5~9	2.69 ± 0.01	1.1 ± 0.1

In these experiments, the waste samples were separated using a manual screening method, and their components were classified. The experimental testing indicators and methods were based on the requirements specified in the National Standards [13–15].

2.3. Monitoring Indicators and Frequency

During the experiment, the indicators of leachate and gas were monitored to assess the stabilization process of the waste. Due to the heterogeneity of the waste in the landfill, many researchers choose to monitor the indicators of the leachate and gas to characterize the stabilization degree of the landfill. Among these indicators, the BOD_5/COD and CO_2/CH_4 in leachate and CO_2/CH_4 in the gas are widely recognized and are significantly related to the stabilization degree of the landfill. When the $BOD_5/COD < 0.1$ and $CO_2/CH_4 < 0.5$, the landfill is considered to have completed the stabilization process [8,16,17]. Therefore, this study observed the indicators of the leachate to characterize the stabilization degree of the waste, and the relevant monitoring indicators are listed in Table 3.

Table 3. The monitoring indexes and frequencies of the gas-water joint regulation system.

Type	Indicators	Method	Frequency
Solid	Compaction density	Sand filling method [18]	Every 15 days
	Settlement	Leveling instrument measurement method	Every 15 days
	Organic compound	Cauterant method	Every 15 days
	VOC	GC-MS	Every 15 days
	Cellulose	DNA colorimetry	Every 15 days
	Hemicellulose	DNA colorimetry	Every 15 days
	Lignin	Klason method [19]	Every 15 days
Liquid	COD	Potassium dichromate oxidation method	Every 15 days
	BOD_5	Dilution and cultivation method	Every 15 days
	Ammonia nitrogen	Distillation titration method	Every 15 days
	Total nitrogen	Alkaline potassium persulfate digestion-UV spectrophotometry	Every 15 days

3. Results and Discussion

3.1. Preliminary Landfill Characterization

The sample, after being crushed and dried, was analyzed through manual screening and mesh screening. The characteristics of waste components from the Landfill are shown in Table 4. The word "mixture" refers to a mixture of components with less than 10 mm particle size.

Table 4. Analysis of waste compositions in a landfill.

Components	Mass Percentage at 2 m Depth (%)	Mass Percentage at 5 m Depth (%)	Mass Percentage at 9 m Depth (%)
Plastic or rubber	36.74	52.22	32.42
Brick, tile, or ceramic	7.03	10.91	6.26
Textile	31.26	12.69	1.04
Metal	6.17	7.38	15.7
Paper	16.6	0.64	3.32
Mixture	2.2	16.16	41.26

Physicochemical Properties

Table 5 shows the physical and chemical characteristics of the waste samples in this experiment.

Table 5. Physical and chemical characteristics of waste in landfill.

Type	Mass Ratio of Waste at 2 m Depth (%)	Mass Ratio of Waste at 5 m Depth (%)	Mass Ratio of Waste at 9 m Depth (%)
Water content	58.94	24.03	17.39
Organic compound	28.74	18.22	9.02
Volatile solid (VS)	44.48	50.14	6.54
BDM	38.78	26.12	3.61
Cellulose (C)	6.04	1.26	0.68
Hemicellulose	39.80	11.18	0.40
Lignin (L)	26.57	24.53	6.38

The results of the waste composition and physicochemical properties show that there are differences in the content of the waste and organic matter at different depths and landfill ages. The landfill age of waste was divided into three parts: less than two years, two to approximately five years, and five to approximately nine years, corresponding to the landfilled depth of 0~2 m, 2~5 m, and 5~9 m, respectively.

3.2. Development of Aeration and Reinjection

Static Respiration Index (SRI) is obtained by observing a change in the oxygen concentration in a closed container with the samples [20]. In contrast, the Dynamic Respiration Index (DRI) is derived by calculating the changes in the concentration of oxygen from the inlet and outlet airflow around samples 12 instantaneous times within 24 h, taking into consideration the solid weight, gas flow rate, and time variation. The DRI represents the highest activity of microorganisms during the four-day observation period [21]. The mathematical expression for the DRI is shown below:

$$\mathrm{DRI} = \sum DRI_{dm}/12 \tag{1}$$

$$DRI_{dm}\ (\mathrm{mgO_2\ kg\ DM^{-1}\ h^{-1}}) = Q \times \theta \times \Delta O_2 \times Vg^{-1} \times 31.98 \times DM^{-1} \times \theta^{-1} \tag{2}$$

where Q is the gas flow rate (measured in L/h); θ is the data acquisition time (s); ΔO_2 is the difference between the oxygen concentration levels at the inlet and outlet gas flow of the reactor (mL/L); Vg is the molar volume of gas, (L/mole); 31.98 is the molecular mass of oxygen, (g/mol); and DM is the dry weight of the sample (kg).

By sampling and analyzing the waste sample in two years of landfill age, the changing trend of waste DRI_{dm} data is shown in Figure 4.

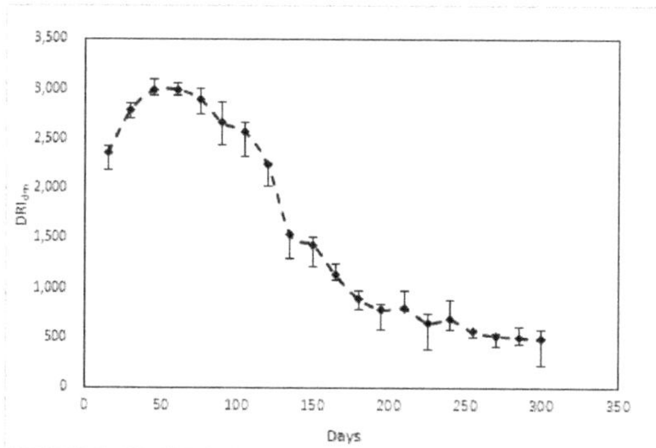

Figure 4. The change of waste DRI_{dm} in the landfill over time.

Researchers found that the Static Respiratory Index (SRI) and Dynamic Respiratory Index (DRI) are both effective in reflecting the biodegradability of solid waste, though the data of both indices change in consistency [22]. Therefore, the ventilation quantity in this experiment is calculated based on the DRI_{dm} value measured from the waste component with the following equation:

$$Q = \frac{DRI_{dm} \times m_i}{(1 - W) \times 0.21 \times 1.429 \times K \times T \times 1000} \quad (3)$$

where Q is the total air volume (measured in L/h); DRI_{dm} is the dynamic respiratory index of the sample, (mg/(kg·h)); m_i is the mass of the sample (kg); W is the moisture content of waste, (%); 0.21 is the volume fraction of oxygen in the air (%); T is the full load operation repair period, (h); 1.429 is the density of oxygen under standard conditions, (g/L); and K is the oxygen utilization rate (%).

According to the DRI_{dm} monitoring results, the peak value for the DRI_{dm} in the past 100 days for the waste anaerobic digestion system was selected as 2985 mg/(kg×h). However, for the 60-day period of the past two years, the value of 1522 mg/(kg×h) was selected as the design value for the 100- to 200-day period based on the 150-day average value from the trial period. The design value for oxygen utilization efficiency was set at 60%.

To maintain the required leachate levels, the amount of backflow is calculated based on the amount of leachate generated from the landfill. This calculation was specified in the "Technical code for leachate treatment of municipal solid waste" CJJ150-2010. The equation is as presented below:

$$Q = \frac{I \times (C_1 A_1 + C_2 A_2 + C_3 A_3)}{1000} \quad (4)$$

where Q is the leachate produced (measured in m^3/d); I is the multi-year average daily rainfall (mm/d); A_1 is the catchment area of the work unit, (m^2), C_1 is the exudation coefficient of the work unit; A_2 is the catchment area of intermediate coverage unit, (m^2); C_2 is the seepage coefficient of intermediate cover unit; A_3 is the catchment area of terminal coverage unit (m^2); and C_3 is exudation coefficient of final covering unit.

The study also considered the maximum water retention of MSW in the landfill for re-injection and daily re-injection of leachate into the reactor with spray irrigation.

The water retention of MSW, the water supply, and the water retention at the landfill were in agreement with the relevant research results, which were determined as 47.19%, 28.75%, and 18.44%, respectively [12]. The minimum area of the re-injection unit (A_{min}) was calculated as follows:

$$A_{min} = \frac{Q}{C \times H} \quad (5)$$

where A_{min} is the minimum area of reinjection unit (m^2/d); Q is the leachate production (m^3/d); C is the water holding capacity of domestic waste (%); and H is the effective landfill depth (m).

In conclusion, the design parameters of the experimental reactors M_1, M_2, and M_3 are shown in Table 6 below:

Table 6. Ventilation and recirculation system parameters.

	Air Volume during 0~100 Days (L/h)	Air Volume during 100~200 Days (L/h)	Air Volume during 200~300 Days (L/h)	Recharge Volume (m^3/d)	Reinjection Area (m^2/d)
M_1	30.78	12.37	7.89	0.196 ± 0.1	0.652
M_2	16.24	8.98	3.40	0.196 ± 0.1	0.652
M_3	0	0	0	0.215 ± 0.1	0.715

3.3. *Effects on Waste and Leachate*

3.3.1. BOD_5 and COD

The variation of leachate in M_1, M_2, and M_3 reactors after 300 days of aerobic remediation testing is shown in the Figures 5 and 6 below:

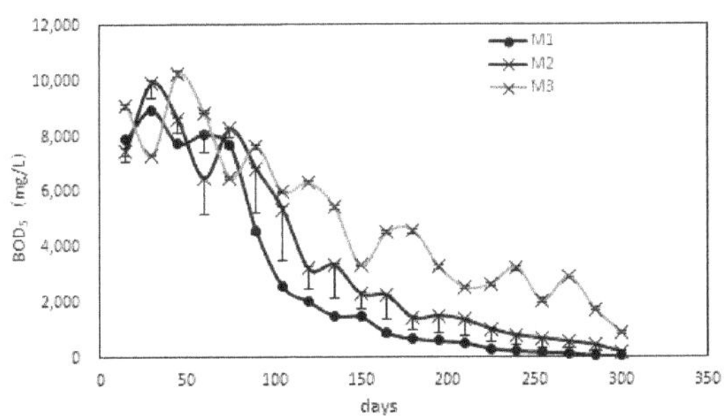

Figure 5. The trend of BOD_5 over time.

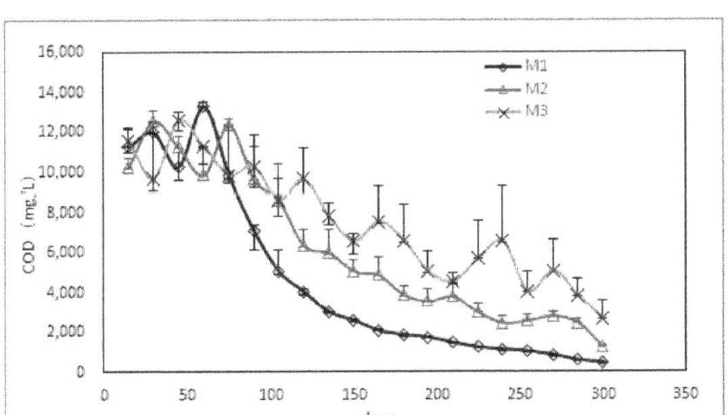

Figure 6. The trend of COD over time.

The results demonstrate that the aerobic-semi aerobic-anaerobic process is highly effective in reducing the BOD_5 and COD values in the leachate of the landfill. The M_1

reactor achieved an outstanding BOD_5 reduction rate of 99.3% and a COD reduction rate of 95.7% when the BOD_5 was reduced to 50 mg/L. The high reduction rate is observed in the M_1 reactor due to the young age of the waste in the reactor, which contained a high concentration of readily decomposable organic matter. The M_2 reactor, which contained waste of intermediate age, achieved a BOD_5 reduction rate of 96.5% and a COD reduction rate of 93.4% when the BOD_5 was reduced to 356.4 mg/L. The M_3 reactor, which contained much older waste, achieved a BOD_5 reduction rate of 67.9% and a COD reduction rate of 49.1% when the BOD_5 was reduced to 1730 mg/L. Although the M_3 reactor had a lower reduction rate, it still achieved a significant reduction in BOD_5 and COD values. The results demonstrate that the aerobic-semi aerobic-anaerobic process has a positive impact on the degradation of organic matter in the leachate, making it a promising method for the treatment of landfill leachate.

3.3.2. BOD5 to COD (B/C) Ratio

The BOD_5 to COD (B/C) ratio, representing the ratio of biodegradable organic carbon in landfill waste, is a crucial indicator of waste stabilization, was shown in Figure 7. The experimental results demonstrated that the B/C values of the M_1 and M_2 reactors are 0.10 and 0.11, respectively, indicating that the waste in these reactors had undergone significant stabilization. The B/C value of the M_3 reactor was relatively higher, at 0.39, suggesting that more time might be required to achieve complete stabilization. These findings are presented in the figure below, which displays the B/C data for each of the three reactors. Overall, these results indicate that the experimental design was successful in achieving waste stabilization.

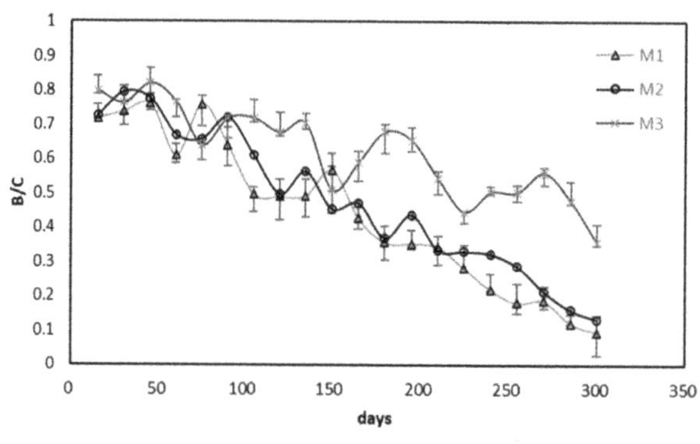

Figure 7. The trend of BOD_5/COD over time.

The high concentration of pollutants in the landfill can be attributed to the fact that most of the waste comes from the surrounding households, producing a significant impact on the landfill's condition. As the landfill's depth increases, the age of the waste gradually increases, with the surface layer containing waste that is no more than one year old, while the waste found in the deeper layers could be older than eight years. The M_1 and M_2 experimental reactors exhibited high concentrations of pollutants because the waste deposited in them was relatively new, ranging from one to five years, and these contained a higher portion of readily decomposable organic matter.

Moreover, the gas-water control technology employed in the experiment proved effective in promoting the decomposition of organic matter in the waste. The B/C ratio for the M_1 and M_2 reactors reached 0.10 and 0.11, respectively, indicating that the waste stabilization process was close to completion. The M_3 reactor also had a lower B/C ratio, indicating that the employed anaerobic technology not only had a positive effect on the decomposition of new waste but also played a role in the anaerobic decomposition of older waste in the landfill. Overall, the experiment demonstrated that effective waste management strategies, such as gas-water control and anaerobic technology, can help to stabilize landfill conditions and promote the decomposition of organic matter in the waste.

3.3.3. NH3-N and TN

Moreover, the experiment demonstrates a significant reduction in NH_3-N and TN concentrations in the M_1, M_2, and M_3 reactors, with corresponding reduction rates of 99.2%, 98.3%, and 88.3% for NH_3-N and 98.8%, 98.5%, and 74.6% for TN, respectively. These results indicate that the aerobic-semi aerobic-anaerobic process effectively aerated and oxidized the leachate in the reactors. The decrease in NH_3-N and TN concentrations is illustrated in the figure below. The reductions in NH_3-N and TN concentrations in the reactors were likely due to the combined effects of the nitrification and denitrification process, as well as the anaerobic degradation of nitrogen-containing organic compounds. Overall, these results suggest that the aerobic-semi aerobic-anaerobic process has the potential to serve as an effective method for the treatment of leachate from landfills. The trend of NH_3-N and TN over time was shown in Figures 8 and 9.

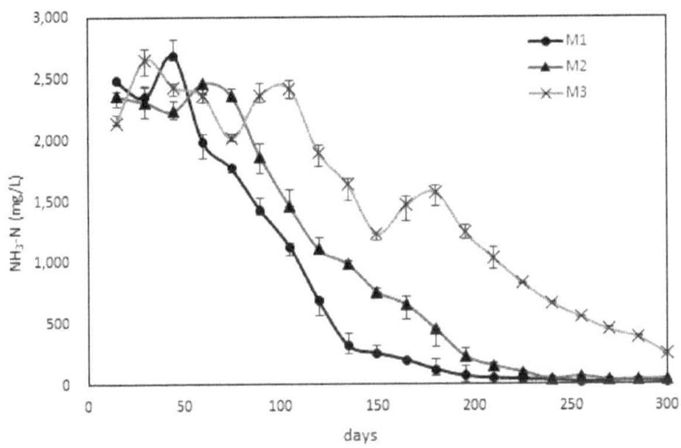

Figure 8. The trend of NH_3-N over time.

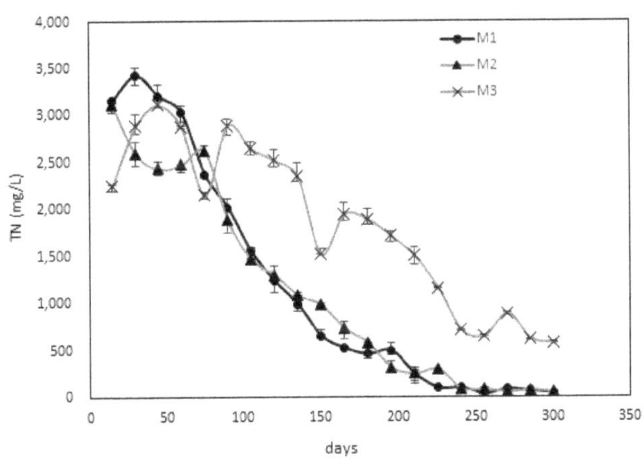

Figure 9. The trend of TN over time.

3.3.4. The Sedimentation Settling Rate Changes

The settling rate is another important parameter that evaluates the stability of the waste landfill. After 300 days of weathering and joint treatment, the settling rate of the M_1 reactor was found to be 260.5 mm, which is significantly higher than that of the M_2 reactor at 162.2 mm. This difference can be attributed to the high proportion of organic matter present in the waste inside the M_1 and M_2 reactors and the positive impact of gas-water joint treatment technology on removing the organic matter from the waste. As a result, the waste in these reactors slowly settles down due to gravity. In contrast, the rate of degradation of the waste in the M_3 reactor is limited; therefore, the settling effect is not obvious. The changes in the settling rate are shown in Figure 10 below, highlighting the significant increase in the settling rate in the M_1 reactor after the treatment.

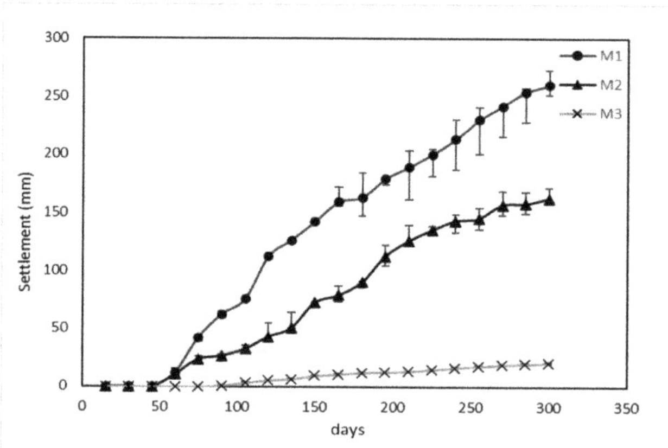

Figure 10. The trend of settling rate over time.

3.4. Discussion

The experiment revealed that the aerobic-semi aerobic-anaerobic reaction has great potential in the rapid and stable treatment of municipal solid waste with different landfill ages. The system proved highly stable and efficient in removing organic components from the waste and leachate. After aeration process, the BOD_5 and COD in the leachate were quickly degraded in the M_1 and M_2 reactors. Additionally, after being treated with aeration, the leachate was re-injected into the M_3 reactor, which further accelerated the reaction, even in anaerobic conditions. The aerobic pre-treatment has a promoting effect on the degradation of fat, carbohydrate, protein, and lignin in the waste components [23]. In addition, aerobic pre-treatment can enhance the methane production ability of landfill. Furthermore, due to the rapid degradation of organic components in the waste during aerobic pre-treatment, the porosity of waste components increases, forming positive feedback and accelerating the rate at which organic matter is decomposed in the waste heap. At the same time, because fresh waste has a high content of easily degraded organic matter, the population and activity of the degradation bacteria in fresh waste are significantly higher than those in aged waste. Moreover, the solubility of organic matter in saturated organic components is significantly higher in fresh waste leachate than in aged waste leachate [24]. Leachate recycling through irrigation is beneficial and increases the population of bacteria in the aged waste, continually promoting the activity of aerobic microorganisms in fresh waste and thus improving the overall metabolic efficiency of the waste.

In this study, various techniques were employed to achieve the effective removal of ammonia and nitrogen from landfill waste. The use of aerobic, semi-aerobic, and anaerobic environments allowed for the improvement of the nitrification-denitrification process, which is crucial for this process. Additionally, joint leachate irrigation technology was utilized to ensure that the leachate remained in the landfill reactors for a sufficient period of time to allow the ammonia and nitrogen to undergo short-term nitrification-denitrification synchronization. This synchronization is vital to effectively remove ammonia and nitrogen from the leachate. Furthermore, researchers conducted quasi-oxygen nitrification-denitrification batch tests and found that different denitrifying bacteria also play a crucial role in the oxygen nitrification-denitrification process [25]. However, in anoxic environments, the removal of nitrogen from the waste is primarily contributed by the anoxic ammonia oxidation bacteria. Thus, optimizing the conditions for both types of bacteria is essential for efficient ammonia and nitrogen removal from landfill waste.

In terms of oxygen utilization efficiency, the indicator of DRI_{dm} was used to depict the oxygen demand of different components in the waste during the gas-water joint regulation process. Combined with leachate back-pumping technology, it promotes the richness and activity of microorganisms in different reaction columns and enables high-efficiency denitrification of short-term nitrification, anoxic ammonification, and aerobic degradation of waste. However, in contrast to the traditional aerobic rehabilitation technology of landfills in China, which uses the BDM or porosity method, stable continuous/intermittent aeration of the landfill is carried out after setting the target value [26,27], and the required oxygen supply for calculation is 92.5 L/h, consumes more oxygen than the two methods mentioned above. Compared with these two methods, the DRI_{dm} indicator can reduce air exposure by 60%, greatly reducing the operation cost.

Moreover, the gas-water joint treatment technology proved to be effective in removing organic matter from the waste, as was evidenced by the high settling rates observed in the M_1 and M_2 reactors. Overall, this study highlights the potential of this technology for sustainable waste management and presents a promising solution for waste treatment in landfills.

4. Conclusions

In-situ aeration technology accelerates the stabilization of MSW landfills through enhancing the degradation of organics, but it still suffers the high energy-consuming and operating costs due to low oxygen utilization efficiency. This research aims to improve oxygen utilization efficiency and therefore improve the designing and engineering of the in-situ aeration process in MSW landfills. Herein three different reactors of M_1, M_2, and M_3 were constructed, with different age wastes in piles, and were treated under aerobic, semi-aerobic, and anaerobic conditions, during which the leachate was recirculated, the dynamic respiratory index was monitored, and gas-water control parameters were optimized at different landfill depths. Furthermore, this study monitored the efficiency of the removal of ammonia and nitrogen from landfill waste, including the use of different environments, leachate irrigation technology, and different types of bacteria. Such techniques are crucial for effective waste management and the protection of the environment. The results show that the simultaneous control of gas and water can significantly improve the rate at which organic matter is removed from the landfill. The COD, BOD, NH_3-N, and TN were all significantly reduced. Compared with the traditional means of aeration, this method can significantly reduce the quantity by more than 60%. Overall, our study has demonstrated the potential benefits of incorporating aerobic pre-treatment and leachate into waste management practices, which can help to improve the efficiency and sustainability of waste treatment processes.

Author Contributions: Methodology, H.Z., T.P. and J.L.; Validation, Y.G.; writing—original draft preparation, J.L.; Writing—review and editing, Y.G. and L.H.; Supervision, G.C. All authors have read and agreed to the published version of the manuscript.

Funding: This research received no external funding.

Institutional Review Board Statement: The study did not require ethical approval.

Informed Consent Statement: Not applicable. The study did not involve humans.

Data Availability Statement: Data is unavailable due to local privacy.

Conflicts of Interest: The authors declare no conflict of interest.

References

1. Read, A.D.; Hudgins, M.; Phillips, P. Perpetual landfilling through aeration of the waste mass; lessons from test cells in Georgia (USA). *Waste Manag.* **2001**, *21*, 617–629. [CrossRef] [PubMed]
2. Slezak, R.; Krzystek, L.; Ledakowicz, S. Degradation of municipal solid waste in simulated landfill bioreactors under aerobic conditions. *Waste Manag.* **2015**, *43*, 293–299. [CrossRef] [PubMed]
3. Liu, L.; Ma, J.; Xue, Q.; Shao, J.; Chen, Y.; Zeng, G. The in-situ aeration in an old landfill in China: Multi-wells optimization method and application. *Waste Manag.* **2018**, *76*, 614–620. [CrossRef]
4. Öncü, G.; Reiser, M.; Kranert, M. Aerobic in situ stabilization of Landfill Konstanz Dorfweiher: Leachate quality after 1 year of operation. *Waste Manag.* **2012**, *32*, 2374–2384. [CrossRef] [PubMed]
5. Heyer, K.U.; Hupe, K.; Ritzkowski, M.; Stegmann, R. Pollutant release and pollutant reduction—Impact of the aeration of landfills. *Waste Manag.* **2005**, *25*, 353–359. [CrossRef] [PubMed]
6. Ritzkowski, M.; Stegmann, R. Landfill aeration worldwide: Concepts, indications and findings. *Waste Manag.* **2012**, *32*, 1411–1419. [CrossRef] [PubMed]
7. Ritzkowski, M.; Heyer, K.-U.; Hupe, K.; Stegmann, R. Long term landfill aeration by means of a wind driven air venting and aeration system. *Twelfth Int. Waste Manag. Landfill Symp.* **2009**, *15*, 879–880.
8. Zhou, H.; Wang, Y.; Wang, P.; Ma, Z. Experimental study on key parameters of oxygen extraction technology in landfill treatment. *Green Technol.* **2015**, *33*, 126–130.
9. Meng, F.; Guo, J.; Liu, Z.; Zhu, W.; Han, J.; Wu, S. Field Testing Study on Design Parameters of Aerobic Stabilization Process in Landfills. *Environ. Sanit. Eng.* **2021**, *29*, 93–98.
10. Kadambala, R.; Powell, J.; Singh, K.; Townsend, T.G. Evaluation of a buried vertical well leachate recirculation system for municipal solid waste landfills. *Waste Manag. Res.* **2016**, *34*, 1300–1306. [CrossRef] [PubMed]
11. Yang, X.; Cao, Z.; Ge, Y.; Dai, X. Design and Practice of Aerobic Stabilized Ventilation System of Landfill Sites. *Green Technol.* **2021**, *20*, 164–167, 171.
12. Li, R. Study on In Situ Aerobic Accelerated Stabilization of Landfill Waste. Ph.D. Thesis, Tsinghua University, Beijing, China, 2013.
13. *GB/T 50123-1999*; Standard for Soil Test Method. China Planning Press: Beijing, China, 1999.
14. *GB/T 23857-2009*; Monitoring and Testing in Municipal Solid Waste Landfill Degradation Treatment. AQSIQ: Beijing, China, 2009.
15. *CJ/T 428-2013*; Detection Methods for the Leachate from Municipal Solid Waste. MOHURD: Beijing, China, 2013.
16. Adani, F.; Lozzi, P.; Genevini, P. Determination of biological stability by oxygen uptake on municipal solid waste and derived products. *Compos. Sci. Util.* **2001**, *9*, 163–178. [CrossRef]
17. Zhang, X. Design on the Landfill Leachate Reduction Control and Treatment Scheme in Tianshui City. Master's Thesis, Lanzhou University, Lanzhou, China, 2012.
18. Xu, B. Quality inspection method of layered compacted subgrade and engineering example analysis. *EDP Sci.* **2021**, *248*, 03068. [CrossRef]
19. Wang, X.; Zhao, J.; Dong, C.; Shen, D.; Xiao, R. Structure characterization and pyrolysis behaviour of Klason lignin from wood. *J. Southeast Univ. Nat. Sci. Ed.* **2014**, *44*, 782–786.
20. Tian, L.; Wang, H.; Yang, Y.; Qu, Z. Remediation of an Informal Landfill Site by In Situ Aeration. *Environ. Eng.* **2019**, *37*, 949–954.
21. Iannotti, D.A.; Pang, T.; Toth, B.L.; Elwell, D.L.; Keener, H.M.; Hoitink, H.A.J. A quantitative respirometric method for monitoring compost stability. *Compos. Sci. Util.* **1993**, *1*, 52–65. [CrossRef]
22. Kelly, R.J. Solid Waste Biodegradation Enhancements and the Evaluation of Analytical Methods Used to Predict Waste Stability. Master's Thesis, Virginia Polytechnic Institute and State University, Blacksburg, VA, USA, 2002.
23. Li, L.; Peng, Y.; Tan, H.; Yang, P.; Ru, L.; Wang, X.; Peng, X. Application status and research progress of in-situ landfill aeration. *China Environ. Sci.* **2021**, *41*, 2725–2736.
24. Xu, H. Large-Scale Model Test and Application of Biochemical, Hydraulic and Mechanical Interaction of High Kitchen Waste. Ph.D. Thesis, Zhejiang University, Zhejiang, China, 2016.
25. Gao, B. Application of solid aerobic biological reaction technology in the treatment of inventory. *Guangdong Chem. Ind.* **2018**, *45*, 217–218, 220.
26. Ni, Z. Regulation Mechanism of MSW "Aerobic-Anaerobic-Aerobic" Three-Stage Bioreactor Landfill. Ph.D. Thesis, Tsinghua University, Beijing, China, 2017.
27. Jiang, C. Study on Acceleration of Waste Stabilization in Semi-Aerobic Landfill. Master's Thesis, Changchun University, Changchun, China, 2020.

processes

MDPI

Article

Nano Zero-Valent Iron (nZVI) Encapsulated with ABS (nZVI/(ABS + EC)) for Sustainable Denitrification Performance and Anti-Aggregation

Fanbin Meng [1,2], Yuning Yang [1], Miao Li [2,*], Qizhi Zhu [1], Bing Qin [1] and Chunpeng Yang [1]

[1] SINOPEC Research Institute of Petroleum Processing Co., Ltd., Beijing 100083, China; mfb1469@163.com (F.M.); qinbing.ripp@sinopec.com (B.Q.); yangchunpeng.ripp@sinopec.com (C.Y.)
[2] School of Environment, Tsinghua University, Beijing 100084, China
* Correspondence: miaoli@tsinghua.edu.cn

Abstract: Aggregation and sharp reactivity decrease are the key problems of using nano zero-valent iron (nZVI) as a potential reaction medium for a permeable reactive barrier (PRB). In this study, nZVI particles encapsulated within an acrylonitrile–butadiene–styrene (ABS) matrix (nZVI/(ABS + EC)) was fabricated, which for the first time successfully simultaneously solved the above problems via accurately regulating the distribution of nZVI particles in the ABS matrix and regulating the contact between nZVI particles and the contaminated aqueous environment. In addition, the size and number of the pores throughout the ABS matrix were first regulated by ethyl cellulose (EC) for the purpose of controlling the contact between nZVI particles and the nitrate contaminant, affording apparent rate constants (k_{obs}) for denitrification performance in the range of 0.0423 to 0.0820 min^{-1}. The remediation of simulated nitrate-contaminated solution by nZVI/(ABS + EC) was suitably described by the first-order kinetics model, with k_{obs} ranging from 0.0423 to 0.2036 min^{-1}, and functional relationship models of k_{obs} with the dosages of EC (d_{EC}) and nZVI (d_{Fe}) during encapsulation were developed for the quantitative regulation of a sustainable denitrification performance. Results revealed that encapsulation prevents the aggregation of nZVI, rendering a sustainable denitrification performance of the material; the denitrification performance was demonstrated to be affected and quantitatively regulated by the encapsulation and application conditions. Using nZVI/(ABS + EC) as the reaction medium for PRB, the pore blocking of PRB can be avoided, and the sustainable remediation performance can be quantitatively regulated and predicted.

Keywords: nano zero-valent iron; permeable reactive barrier; denitrification performance; aggregation; reactivity decrease

1. Introduction

Nitrate is a pervasive and typical groundwater contaminant [1] originating from sources like industrial and domestic waste [2], extensive use of fertilizers [3], and emissions of animal dung [4]. Moreover, the consumption of nitrate-contaminated groundwater can increase the risk of gastric and esophageal cancers and lead to methemoglobinemia in babies [5,6].

As the key technology for the in situ remediation of nitrate-contaminated groundwater, currently, permeable reactive barrier (PRB) remediation systems mainly use highly reductive, cost-effective zero-valent iron (ZVI) as the reaction medium [7,8]. Owing to its high specific surface area, nZVI exhibits a high reactivity and contamination removal rate. A kinetics study revealed that the reduction rate of nitrate by nZVI is 10 to 100 times more rapid than that of micro-scale iron particles [9–11]. Hence, PRB using nZVI as the reaction medium exhibits an advantage for the removal rate of the contaminant and elimination of the pollutants at high concentrations. In view of the rapid development of nanotechnology and considerable potential advantages of nZVI, experiments and applications using nZVI as the reaction medium for PRB have been performed by several researchers [12–14].

However, aggregation [15,16] and poor reactivity persistence [17] are the bottleneck issues in the use of nZVI as a reaction medium for PRB [18,19]. Hence, several studies [20,21] have been extensively conducted to overcome these bottleneck issues. The encapsulation of nZVI particles in a porous matrix is a typical approach to immobilize nZVI particles and ultimately prevent aggregation [22]. Calcium alginate beads constitute one of the typical encapsulation materials, and these beads have been used to immobilize nZVI particles [23,24]. Encapsulated nZVI particles can contact and react with contaminants via holes throughout the calcium alginate beads. In addition to calcium alginate, polystyrene (PS) [25], chitosan [26], and carbon spheres [27,28] have been used to prevent the aggregation of nZVI particles.

To increase the reactivity persistence of nZVI, several strategies [20,21] have been developed to extend its reactivity period. The most sophisticated method to improve the longevity of nZVI is to emulsify nZVI particles with vegetable oil and other additives to decrease their corrosion rate [29]. The emulsification of nZVI considerably increases its longevity by restraining the contact between nZVI particles and the contaminant [21]; however, owing to its good flow behavior, nZVI emulsified with vegetable oil is only suitable for in situ remediation and not for PRB remediation. Complicated hydrogeology may have resulted in the destabilization of the nZVI and vegetable oil suspension, ultimately leading to aggregation. Generally, previous studies have focused on solving either the aggregation or poor reactivity persistence; however, few studies on synchronously solving both aggregation and poor reactivity persistence have been conducted, and studies on the quantitative regulation of the nZVI reactivity persistence have not been conducted.

Synchronously solving both aggregation and poor reactivity persistence constitutes the core issue for using nZVI as the reactive medium for PRB; therefore, the key technology is to synchronously implement the immobilization of nZVI particles and control the contact between the nZVI particles and contaminant. Encapsulation is typically employed to stabilize active ingredients in drug delivery [30–33], pesticides [34–37], self-healing [38], and food [39–41]. However, current studies [42] devoted to the use of nZVI as an alternative reactive media for PRB have mostly focused on the use of encapsulation for immobilizing nZVI to prevent aggregation. Encapsulation has been rarely applied to improve the longevity of nZVI, and attempts have not been made to employ encapsulation to simultaneously reduce aggregation and prevent the sharp reactivity loss of nZVI to the best of our knowledge. The ability of encapsulation to immobilize nZVI and prevent aggregation has been demonstrated by previous studies. Previous studies have reported that encapsulation can enable the sustainable release of reactive nZVI and ultimately maintain its continuous reactivity. Hence, encapsulation is feasible to simultaneously prevent aggregation and increase the reactivity persistence of nZVI.

However, to the best of our knowledge, studies have not been conducted to employ the encapsulation for the synchronous solution of aggregation and sharp reactivity decrease in nZVI and to quantitatively regulate the reactivity persistence of nZVI. Hence, this study explores the encapsulation of nZVI particles into an acrylonitrile–butadiene–styrene (ABS) matrix to immobilize nZVI particles. In addition, different dosages of ethyl cellulose (EC) were used to modify the pores throughout the ABS matrix to regulate the size and number of holes and further control the contact between nZVI particles and the contaminant. Moreover, the apparent rate coefficient (k_{obs}) for nitrate removal using encapsulated nZVI was calculated by numerical simulations by utilizing different dosages of EC and Fe^0 based on batch experiments. In addition, effects of the morphology, structure, and denitrification conditions on the denitrification performance of nZVI encapsulated in ABS were thoroughly investigated.

2. Experimental

2.1. Materials and Methods

ABS (industrial grade, M_w = 10,000–12,000) was provided by Langfang Juxin Plastic Factory (Langfang, China). Both ethyl cellulose (EC) and polyvinyl pyriloxanone (PVP, K30) were supplied by Sinopharm Chemical Reagent Co., Ltd. (Beijing, China). Ferrous sulfate ($FeSO_4 \bullet 7H_2O$, >99%) was purchased from Fuchen (Tianjin) Chemical Reagent Co., Ltd. (Tianjin, China) Methylene dichloride (CH_2Cl_2, >98%), ethanol (CH_3CH_2OH, >99.7%), and n-heptane were purchased from Beijing Chemical Works (Beijing, China). Sodium borohydride ($NaBH_4$, >99%) was producted by Shanghai Aladdin Biochemical Technology Co., Ltd. (Shanghai, China). Polydimethylsiloxane (PDMS, commercial grade, 3000 cSt) were obtained from Dow Coring XIAMETER® (DeLand, FL, USA). All reagents were used as received. Deionized water was used throughout this research and was deoxygenated before use.

In this study, reagent-grade $NaNO_3$, which was purchased from Sinopharm Chemical Reagent Co., Ltd. (Beijing, China), was chosen as the target compound. The simulated pollutant solution with different NO_3^--N concentrations were prepared by dissolving specified amounts of $NaNO_3$ in deoxygenated water.

To monitor the denitrification performance of the following samples, 0.5 mL aliquots were extracted at various reaction times. Then, they were diluted to 25.0 mL to analyze the NO_3^--N concentrations using ultraviolet spectrophotometry (HACH DR6000, HACH, Loveland, CO, USA) at the absorbance wavelengths of 220 and 275 nm.

2.2. The nZVI Preparation

The synthesis of nZVI was carried out by a classical liquid phase reduction method. At room temperature (25 °C), 0.50 g of PVP was added to a $FeSO_4 \bullet 7H_2O$ solution (50 mL, 0.01 mol/L) and stirred by a mechanical agitator at a rate of 200 r/min. After 20 min, the PVP was completely dissolved, and then 50 mL of $NaBH_4$ solution (0.03 mol/L) was poured into the above mixed solution of $FeSO_4 \bullet 7H_2O$ and PVP. After 10 min, the generated Nzvi particles were collected via magnetic separation. Finally, the obtained Nzvi particles were washed three times with 100 mL of ethanol, and then dried at 60 °C using a vacuum drying oven (DHG-9005, Yi Heng, Shanghai, China).

2.3. Encapsulation of Nzvi Particles

First, the shell material solution was prepared by dissolving 0.25 g of ABS and the desired weight of EC into 30 mL of methylene dichloride in a 100 mL three-neck round-bottom flask under 200 r/min stirring. After that, the synthesized Nzvi particles were added into the above shell material solution. After 10 min, the nZVI particles were successfully suspended in the shell material solution, and then 30 mL of PDMS was added dropwise into the above suspension to induce the coacervation of ABS and EC. After titration, the mixture in the three-neck round-bottom flask was poured into 250 mL of n-heptane to solidify the shell, and the obtained particles were also collected via magnetic separation. The collected particles were washed three times with 100 mL of n-heptane, and then soaked in 50 mL of ethanol for 24 h to remove EC. Finally, the particles were separated from the ethanol and dried at 60 °C in a vacuum drying oven for 6 h, and the obtained sample was named nZVI/(ABS + EC). In addition, the parameter conditions and nomenclature for the samples prepared in this study are shown in Table 1.

Table 1. Nomenclature for nZVI/(ABS + EC) samples fabricated under various encapsulation conditions.

Sample No.	m (ABS)	m (EC)	m (nZVI)	Nomenclature
1	0.25	0.15	0.15	nZVI/(ABS + EC) - (0.6–0.6)
2	0.25	0.15	0.20	nZVI/(ABS + EC) - (0.6–0.8)
3	0.25	0.15	0.25	nZVI/(ABS + EC) - (0.6–1.0)
4	0.25	0.15	0.50	nZVI/(ABS + EC) - (0.6–2.0)
5	0.25	0.00	0.25	nZVI/ABS
6	0.25	0.02	0.25	nZVI/(ABS + EC) - (0.08–1.0)
7	0.25	0.05	0.25	nZVI/(ABS + EC) - (0.2–1.0)
8	0.25	0.10	0.25	nZVI/(ABS + EC) - (0.4–1.0)
9	0.25	0.20	0.25	nZVI/(ABS + EC) - (0.8–1.0)
10	0.25	0	0	ABS

2.4. Characterization of the nZVI/(ABS + EC)

SEM:

Scanning electron microscopy (SEM, ZEISS Merlin, Oberkochen, Germany) was used to observe the morphology of the ABS-encapsulated nZVI, the virgin nZVI/(ABS + EC) - (0.6–0.8), and nZVI/(ABS + EC) - (0.6–0.8) after the denitrification reaction. Also, morphology changes in the ABS and EC composites after immersion in ethanol were observed by SEM to verify the fabrication mechanism of nZVI/(ABS + EC). Finally, nZVI/(ABS + EC) - (0.6–0.8) was deironized after the deironization by immersion in the hydrochloric acid solution (0.1 mol/L, 50 mL) for 48 h and then the matrix morphology of nZVI/(ABS + EC) was observed by SEM.

BET:

The specific surface area, total pore volume and porosity of the nZVI/ABS, nZVI/(ABS + EC) without alcoholic impregnation, and nZVI/(ABS + EC) after alcoholic impregnation were measured by a BET-surface area analyzer (Quantachrome Instruments, America), and calculated by the Brunauer–Emmett–Teller (BET) and Barrett–Joyner–Halenda (BJH) models from the N_2 adsorption isotherms.

Batch experiment:

Batch experiments were carried out to determine the influence of preparation and denitrification conditions on the denitrification performance of nZVI/(ABS + EC). A comparison experiment using nZVI and ABS as the denitrification agent was also conducted. All batch experiments were conducted in 50 mL glass bottles under an anaerobic environment, and the reaction temperature was controlled by a water bath (HH-8 Kexi, Kexi Co. Ltd., Beijing, China). To create an anaerobic environment for our denitrification experiments, upon addition of nZVI/(ABS + EC) to glass bottles containing the simulated pollutant solution, the bottles were promptly sealed with aluminum caps and rubber stoppers to prevent air ingress. Sampling at predetermined time intervals was conducted by penetrating the rubber stopper with a syringe to extract 0.5 mL of sample, ensuring the maintenance of anaerobic conditions throughout the nitrate removal process.

The effects of EC and nZVI dosage during preparation on the denitrification performance of encapsulated nZVI were studied. The remediation of a solution of 50 mg/L NO_3^--N (25 mL) was conducted using 0.1 g of nZVI/(ABS + EC) fabricated with various EC and nZVI dosages at 25 °C and an unregulated initial pH.

The factors for the NO_3^--N removal by nZVI/(ABS + EC) were investigated by remediating a NO_3^--N solution (25 mL) at 25 °C, using nZVI/(ABS + EC) - (0.6–0.8) as the reductant to remove nitrate.

(1) The initial pH of the system

Batch experiments were conducted at pH values of 3, 5, 7, 9, and 11 with an nZVI/(ABS + EC) - (0.6–0.8) dosage of 4 g/L and an initial NO_3^--N concentration of 50 mg/L.

(2) The initial NO_3^--N concentration

Batch experiments were conducted at initial NO_3^--N concentrations (c_0) of 15 mg/L, 25 mg/L, 50 mg/L, 75 mg/L, and 100 mg/L with an nZVI/(ABS + EC) - (0.6–0.8) dosage of 4 g/L without controlling the initial pH.

(3) The reductant concentration

Batch experiments were conducted at initial reductant concentrations of 2 g/L, 4 g/L, 6 g/L, and 8 g/L with an NO_3^--N concentration of 50 mg/L and an uncontrolled initial pH.

In this study, based on the results of the above batch experiments, the denitrification kinetics by nZVI/(ABS + EC) under various preparation conditions and denitrification conditions were investigated by developing an *n*th-order reaction model (Equation (1)).

$$r = -\frac{d[NO_3^-]}{dt} = k_{obs}[NO_3^-]^n \quad (1)$$

where k_{obs} is the observed *n*th-order reaction rate constant. The total duration of each set of experiments was extended to 48 h to comprehensively assess the nitrate removal capability of nZVI/(ABS + EC) over a broad range of operational conditions. However, detailed observations and data analysis focused on the initial 24 h, during which the most significant nitrate removal was observed, highlighting the rapid effectiveness of the nZVI/(ABS + EC) system in reducing NO_3^--N concentrations.

3. Results and Discussion

3.1. nZVI and Matrix Modification Mechanism

The SEM images in Figure 1 show that the obtained Fe^0 particles were 70 nm and spherical in shape. Obvious aggregation of nZVI particles appeared on the SEM image. What is worse, the aggregation of nZVI particles deteriorated during the denitrification process for the Fe^0 corrosion reaction. ABS was used to encapsulate water-soluble core materials (such as oxalic acid [43], potassium persulate [44]) for sustainable release performance. Some small holes appeared on the surface of the ABS matrix, which can significantly affect the nZVI release rate. Moreover, the surface appearance of the ABS and EC composite matrix was similar to that of ABS (Figure 1b,c).

However, both the nZVI encapsulated with ABS and the composite of ABS and EC exhibited very slow denitrification rates. The primary analysis suggests that the size and number of pores influenced the denitrification rate, and these pores may have been produced by the ethanol impregnation process of the ABS and EC composite (Figure 2). After impregnation, more and larger holes appeared on the surface of the ABS and EC composite matrix (Figure 2d). It can be inferred that the size and number of the pores was controlled by the EC content of the composite.

3.2. Morphology Observation

After alcoholic impregnation, many pores were exposed on the matrix surface, and a mass of catenaries constituted by nZVI particles appeared on the pore walls (Figure 3a,b). The nZVI particles spread on the pore wall could limit particle movement and ultimately prevent nZVI nanoparticle aggregation. After deironization via immersion in dilute HCl solution, many more pores were observed in the nZVI/(ABS + EC) - (0.6–1.0) (Figure 3c), which may have been caused by the removal of nZVI particles embedded in nZVI/(ABS + EC) - (0.6–1.0). Also, the morphology of the nZVI/(ABS + EC) - (0.6–1.0) after denitrification was observed by SEM, which showed that a number of irregularly-shaped particles were attached to the surface and pore wall (Figure 3d), which could impede the contact of the nZVI particles and contaminants. Therefore, it can be inferred from the obtained SEM images that a large number of pores was produced by the fabrication method, and the nZVI particles were successfully supported on the pore walls or were embedded within the matrix. However, the deposition of iron oxides generated from the denitrification reaction may adversely affect the denitrification rate.

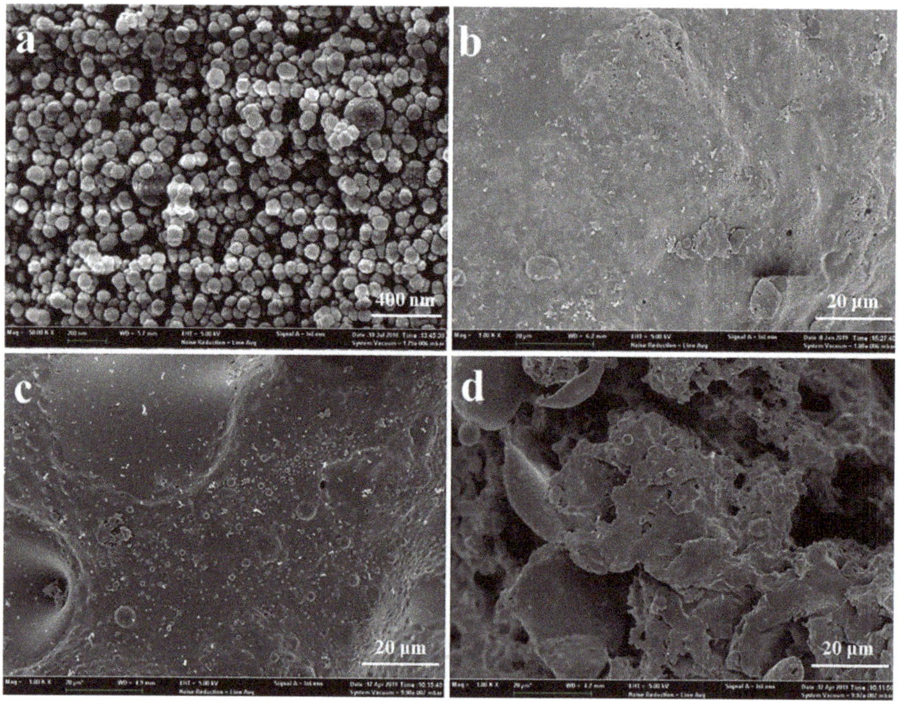

Figure 1. SEM images of nZVI (**a**), ABS matrix (**b**), the matrix of the ABS and EC composite (**c**), and the matrix of the ABS and EC composite after ethanol impregnation (**d**).

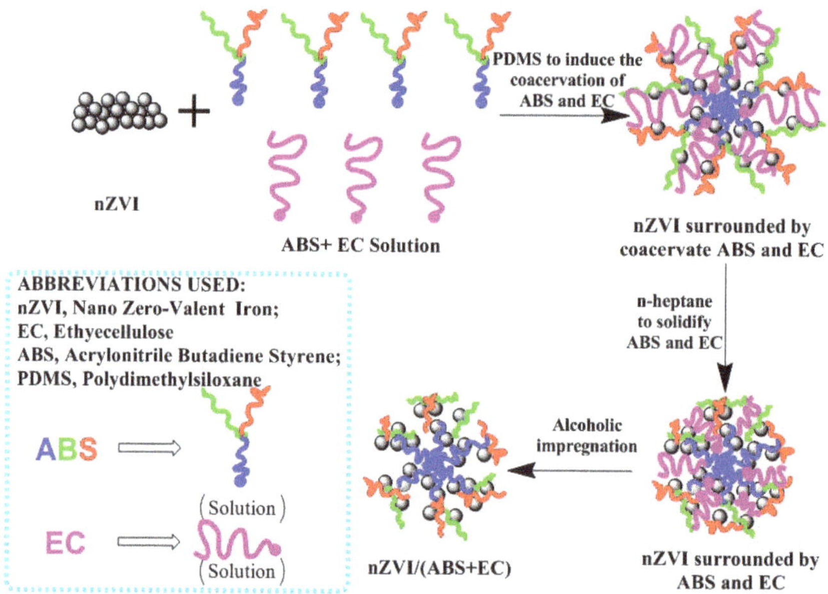

Figure 2. The encapsulation mechanism of nZVI in the ABS matrix via coacervation induced by PDMS.

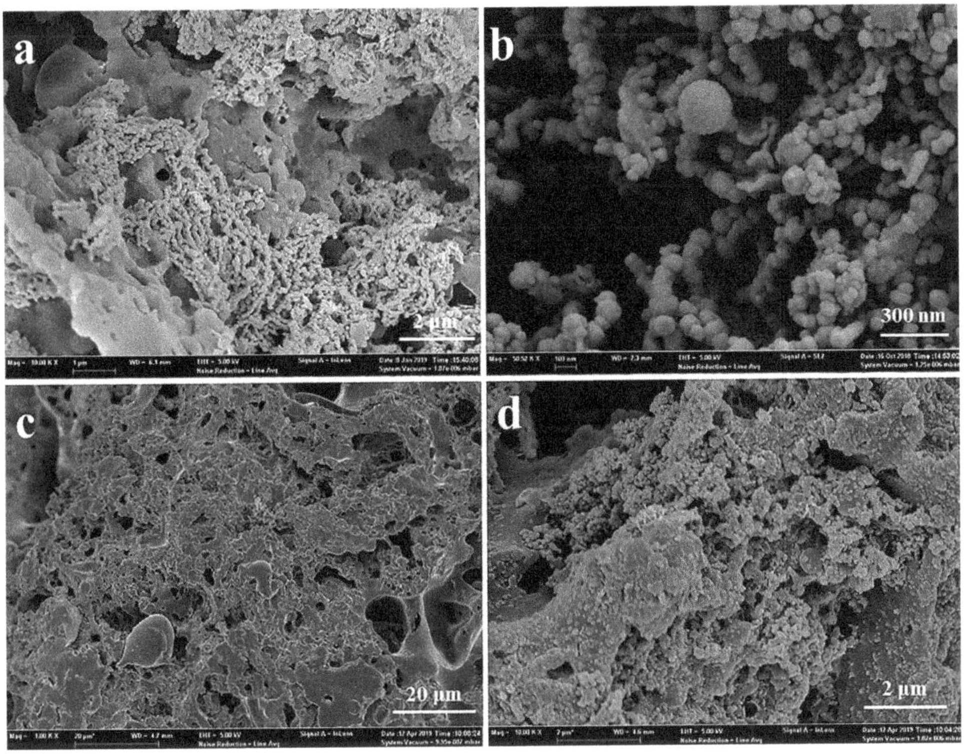

Figure 3. SEM images of nZVI/(ABS + EC) - (0.6–1.0) (**a**,**b**), nZVI/(ABS + EC) - (0.6–1.0) after the deironization (**c**), and nZVI/(ABS + EC) - (0.6–1.0) (**a**,**b**) after denitrification (**d**).

3.3. BET Test

All N_2 adsorption/desorption isotherms of nZVI/ABS, nZVI/(ABS + EC) without alcoholic impregnation, and nZVI/(ABS + EC) after alcoholic impregnation displayed a typical type IV isotherm (IUPAC classification) with an H3 hysteresis loop (Figure 4). These results indicate the mesoporous and macroporous structures of all samples. This is also supported by the pore size distribution (Figure 4 inset image), which shows that the nZVI/(ABS + EC) after alcoholic impregnation was dominated by many small holes with a diameter greater than 15 nm. In addition, the pore volumes of nZVI/ABS, nZVI/(ABS + EC) without alcoholic impregnation, and nZVI/(ABS + EC) after alcoholic impregnation were 0.074, 0.071, and 0.130 cc/g, respectively. The obtained BET results confirm that the holes in the matrix were formed due to alcoholic impregnation, which agrees with the SEM observations. Therefore, it can be inferred that the alcoholic impregnation process can be used to achieve the sustained release of nZVI encapsulated by an ABS and EC composite.

3.4. Effects of EC Dosage

The ABS particles showed almost no removal effect on nitrate, moreover, the denitrification rates of all encapsulated nZVI samples were notably lower than that of bare nZVI (Figure 5a), indicating that the ABS alone had negligible effects on nitrate reduction, but the encapsulation with ABS enabled sustainable denitrification. The maximum adsorption capacity of nZVI prepared in this study for nitrate was as high as 142.952 mg/g (m_N/m_{nZVI}). However, compared to nZVI/(ABS + EC) samples, nZVI exhibited extremely fast denitrification rates (Figure 5a), indicating that encapsulation enabled sustainable denitrification. The obtained results (Table 2) indicate a first-order dependence of denitrification on EC dosage during encapsulation. Moreover, k_{obs} increased as d_{EC} increased, which was attributed to a higher number of holes in the matrix (Figure 5b). The number of holes was affected by the EC dosage during encapsulation, which further influenced the denitrification rate of the encapsulated nZVI. For denitrification at an initial NO_3^--N concentration of 50 mg/L, with a dosage of 0.10 g encapsulated nZVI, the functional relationship between k_{obs} and d_{EC} is given as follows (Equation (2)):

$$k_{obs} = -0.7463 \times {d_{EC}}^2 + 0.3898 \times d_{EC} + 0.0342 \quad (2)$$

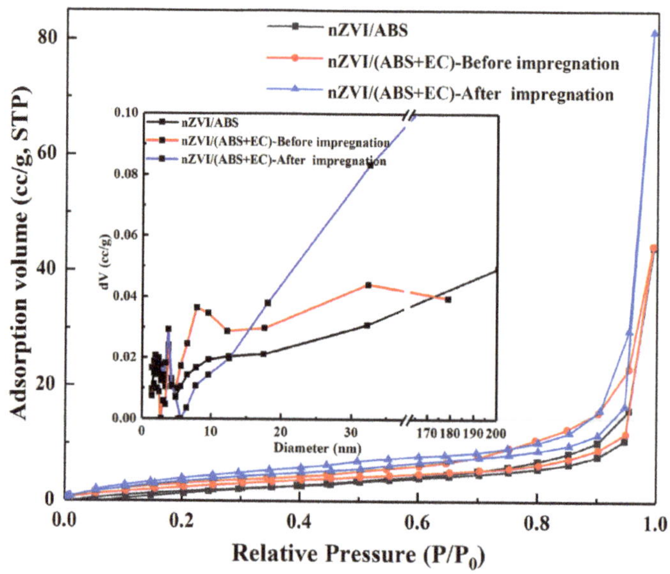

Figure 4. The N_2 adsorption/desorption isotherm and BJH (Barrett–Joyner–Halenda) pore size distribution (inset) of nZVI/ABS, nZVI/(ABS + EC) without alcoholic impregnation, and nZVI/(ABS + EC) after alcoholic impregnation.

Table 2. The k_{obs} of samples prepared with different EC dosages during encapsulation.

Sample	d_{EC}	k_{obs}/min^{-1}	R^2
nZVI/(ABS + EC) - (0.08–1.0)	0.02	0.0423	0.9500
nZVI/(ABS + EC) - (0.2–1.0)	0.05	0.0507	0.9304
nZVI/(ABS + EC) - (0.4–1.0)	0.10	0.0659	0.9385
nZVI/(ABS + EC) - (0.6–1.0)	0.15	0.0763	0.8805
nZVI/(ABS + EC) - (0.8–1.0)	0.20	0.0820	0.8695

This outcome reinforces our hypothesis that the observed denitrification is primarily attributable to the activity of nZVI. Therefore, it can be concluded that the EC dosage during encapsulation was an important parameter influencing the denitrification rate when encapsulated nZVI was used to reduce nitrate. The sustainable denitrification rate can be quantitatively regulated by the EC dosage during encapsulation.

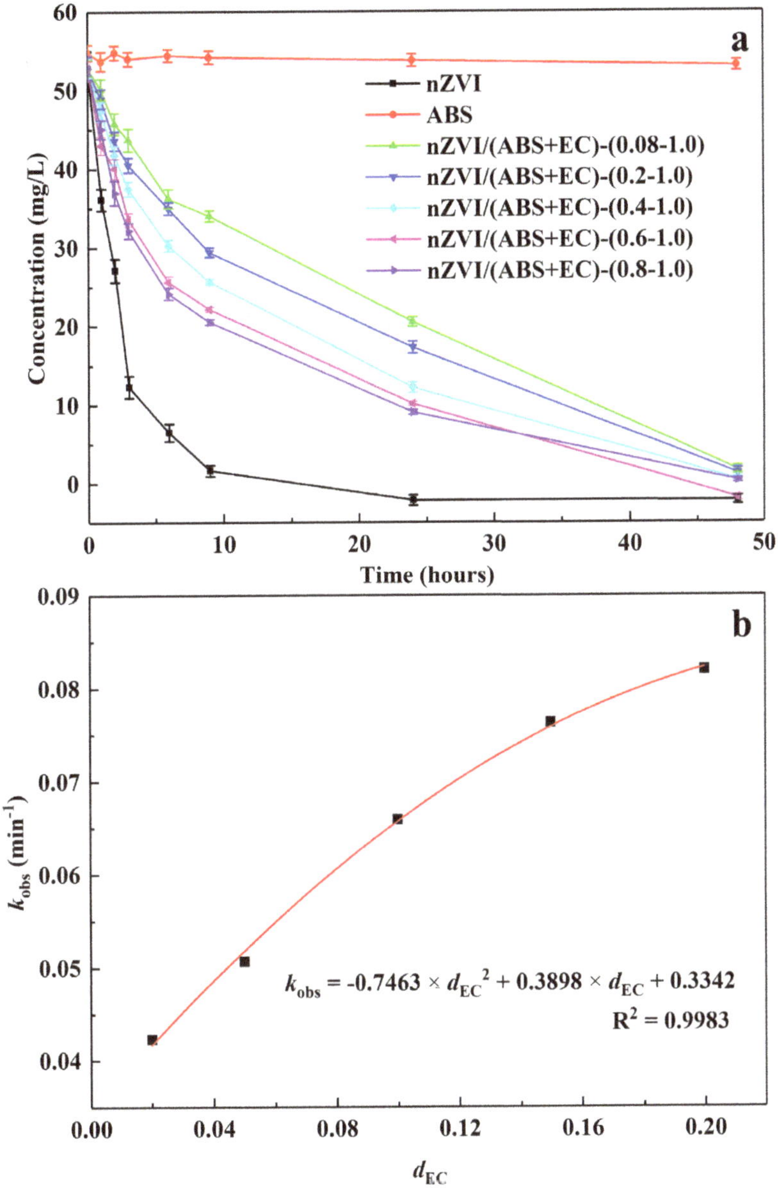

Figure 5. The denitrification performance (**a**) and k_{obs} (**b**) of nZVI and samples prepared under different EC dosages during encapsulation.

3.5. Effects of nZVI Dosage

The encapsulated nZVI samples had an obvious sustainable denitrification effect, and the denitrification rate of encapsulated nZVI was affected by the nZVI dosage (d_{Fe}) during encapsulation (Figure 6a). A linear model closely fit the data (Table 3, $R^2 \geq 0.9902$ in all cases), demonstrating that the reaction was first order with respect to the nZVI dosage during encapsulation. Also, k_{obs} increased with d_{Fe} (Figure 6b), and the calculation of k_{obs} can be expressed by the following equation (Equation (3)):

$$k_{obs} = -0.3128 \times d_{Fe}^2 + 0.5361 \times d_{Fe} + 0.014 \quad (3)$$

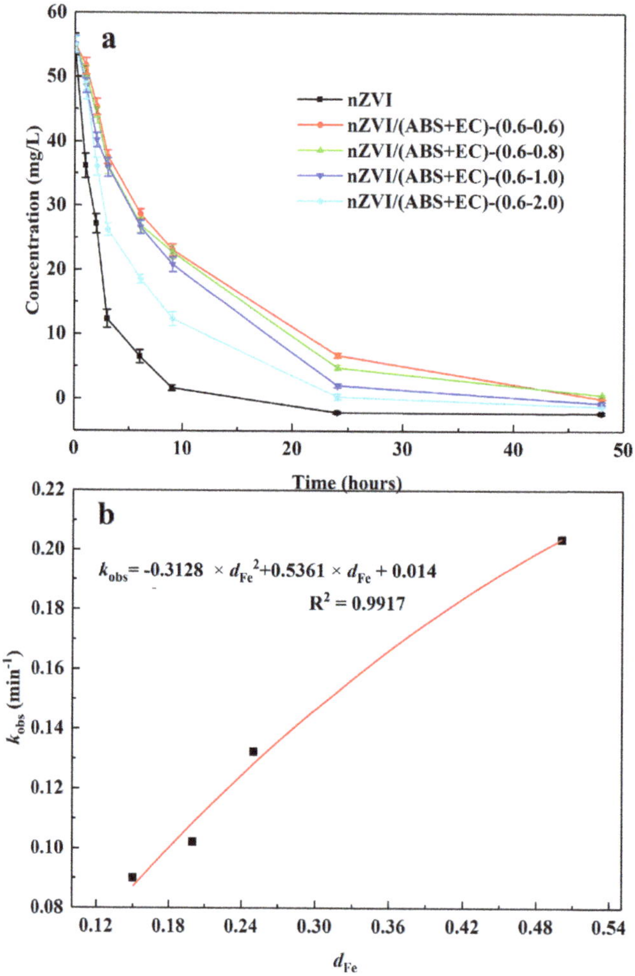

Figure 6. The denitrification performance (**a**) and k_{obs} (**b**) of nZVI and samples prepared under different nZVI dosages during encapsulation.

Table 3. The k_{obs} of samples prepared under different nZVI dosages during encapsulation.

Sample	d_{Fe}	k_{obs}/min^{-1}	R^2
nZVI/(ABS + EC) - (0.6–0.6)	0.15	0.0901	0.9895
nZVI/(ABS + EC) - (0.6–0.8)	0.20	0.1022	0.9939
nZVI/(ABS + EC) - (0.6–1.0)	0.25	0.1323	0.9914
nZVI/(ABS + EC) - (0.6–2.0)	0.50	0.2036	0.9902

The above results indicate that, in addition to the EC dosage during encapsulation, the nZVI dosage significantly influenced the denitrification rate. Thus, adjusting the nZVI dosage during encapsulation is a feasible method to regulate the sustainable denitrification rate.

3.6. Effects of pH

When the initial pH ranged from 3 to 11, nZVI/(ABS + EC) - (0.6–0.8) presented a fast denitrification rate with some differences (Figure 7), and the effect of pH on the denitrification rate was highly variable. The fastest denitrification rates were observed at initial pH values of 3 and 9. The slowest denitrification rate was observed at the early stage of the reaction at an initial pH of 5, and also during the later reaction stage at an initial pH of 11. To explain the above phenomena, it is necessary to examine the denitrification mechanisms of Fe^0 proposed by Xu [45].

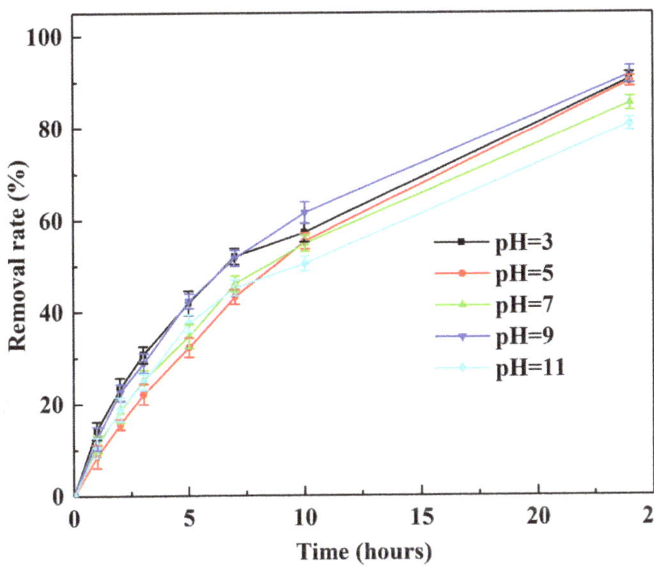

Figure 7. The influence of pH on the denitrification performance of nZVI/(ABS + EC) - (0.6–0.8).

Primarily, the reduction of nitrate occurs due to an electron provided by Fe^0 (Equations (4) and (5)).

$$Fe^0 \rightarrow Fe^{2+} + 2e^- \quad (4)$$

$$NO_3^- + 10H^+ + 8e^- \rightarrow NH_4^+ + 3H_2O \quad (5)$$

When the concentration of H^+ in the system was low, the nitrate reduction rate decreased, accompanied by the generation of Fe_3O_4 (Equations (6) and (7)).

$$3Fe^0 + 3H_2O \rightarrow Fe_3O_4 + 8H^+ + 8e^- \quad (6)$$

$$3Fe^{2+} + 4H_2O + 8e^- \rightarrow Fe_3O_4 + 8H^+ + 2e^- \quad (7)$$

The corrosion of the iron surface would supply electrons for nitrate reduction. The generated Fe_3O_4 is a good conductor for electron transfer and eventually increased the nitrate removal rate. However, corrosion and Fe_3O_4 deposition on the iron surface were significantly influenced by the system pH. An acidic pH accelerated the iron surface corrosion and increased the nitrate removal rate at a pH of 3. When the system was under weakly basic conditions, Fe_3O_4 was easily generated, and the denitrification rate was also fast at a pH of 9. However, to explain the nitrate removal performance of the encapsulated nZVI between pH values from 5 to 7, the contribution of corrosion and Fe_3O_4 deposition to nitrate removal should be quantitatively evaluated. The above results indicate that encapsulated nZVI could rapidly remove nitrate between pH values from 3 to 11, and the pH of the reducing system was not significant, but it had a complicated effect on the denitrification rate.

3.7. Effects of Initial Nitrate Concentration

Previous studies have confirmed that the initial NO_3^--N concentration is intimately related to the denitrification rate of nZVI. However, Wang et al. [46] reported that the denitrification rate of nZVI decreased with the initial NO_3^--N concentration, but Yang et al. [10] reported the opposite. Moreover, encapsulated nZVI showed complicated denitrification performance at different initial NO_3^--N concentrations (Figure 8a). In this study, the nitrate reduction by nZVI/(ABS + EC) - (0.6–0.8) at different initial NO_3^--N concentrations is described by a first-order kinetic model (Table 4). However, the impact of the initial NO_3^--N concentration on k_{obs} was not monotonic (Figure 8b), and the reported relevant literatures has not reached a consensus on this issue [10,47].

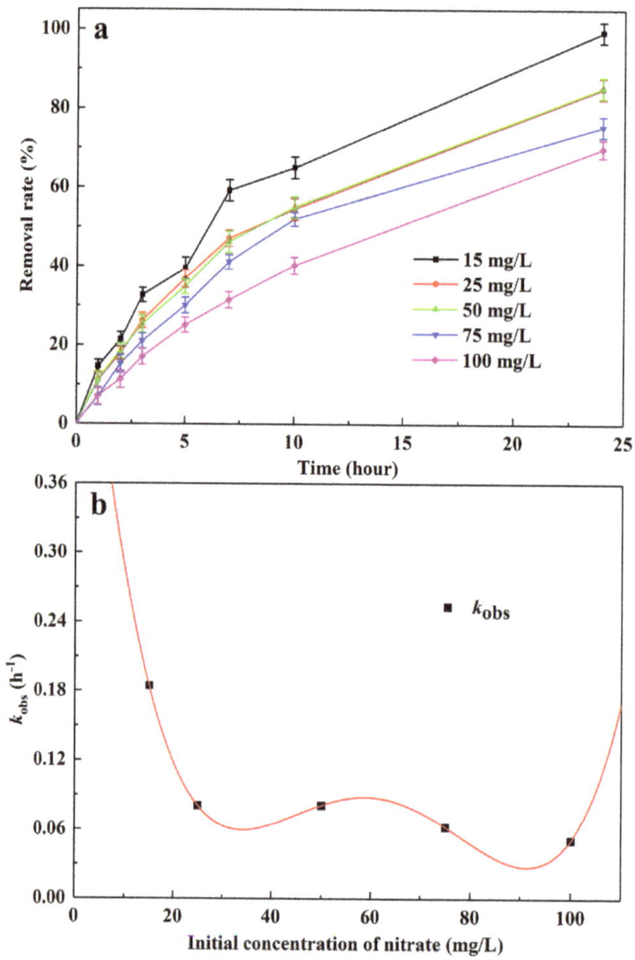

Figure 8. The denitrification performance (**a**) and k_{obs} (**b**) of nZVI/(ABS + EC) - (0.6–0.8) under different initial NO_3^--N concentrations.

Table 4. The k_{obs} (b) of nZVI/(ABS + EC) - (0.6–0.8) under different initial NO_3^--N concentrations.

No.	Initial Nitrate Concentration/(mg/L)	k_{obs}	R^2
1	15	0.1844	0.9296
2	25	0.0807	0.9938
3	50	0.0809	0.9963
4	75	0.0623	0.9759
5	100	0.0509	0.9960

In this study, k_{obs} decreased as the c_0 of NO_3^--N increased from 15 to 25 mg/L, but it increased when c_0 was greater than 100 mg/L. This phenomenon may be explained as follows. When the NO_3^--N concentration was low, the concentration of encapsulated nZVI was relatively high, and adsorption played a dominant role in the removal of NO_3^--N. Therefore, k_{obs} decreased as c_0 increased at low initial NO_3^--N concentrations. However, when the NO_3^--N concentration was high, the concentration of the encapsulated nZVI was relatively low, and reduction played a dominant role in the removal of NO_3^--N, accompanied by the generation of Fe_3O_4 and Fe^{2+} (Figure 9) [45,48,49]. Moreover, Fe^{2+} and Fe_3O_4 had prominent promotion effects on nitrate reduction and increasing the initial NO_3^--N concentration accelerated the generation of Fe_3O_4 and Fe^{2+} [45], which eventually reduced nitrate. The change in the NO_3^--N removal rate was complicated when the NO_3^--N concentration ranged from 25 to 100 mg/L, possibly due to adsorption and reduction processes. The obtained results indicate that adsorption and reduction were both involved during NO_3^--N removal, and the initial NO_3^--N concentration influenced the denitrification rate of encapsulated nZVI by adsorption and reduction.

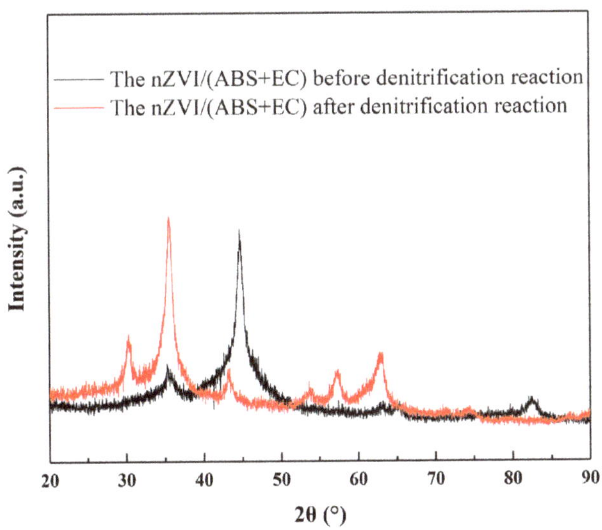

Figure 9. The XRD spectroscopy of the nZVI/(ABS + EC) - (0.6–0.8) before and after denitrification reaction under the initial NO_3^--N concentrations of 100 mg/L.

The complexity of this relationship particularly manifests within the intermediate concentration range (25 to 100 mg/L), where neither adsorption nor reduction can be solely credited for the observed denitrification performance. In conclusion, while our study has preliminarily shed light on the intricate dynamics between the initial NO_3^--N concentration and the denitrification rate of encapsulated nZVI, it also underscores the necessity for further research. To elucidate the observed complexities, a thorough and quantitative investigation of the denitrification mechanisms, taking into account factors such as pH, nZVI/(ABS + EC) concentration, and encapsulation effects, is imperative.

3.8. *Effects of Reductant Concentration*

The reduction of nitrate using nZVI/(ABS + EC) - (0.6–0.8) was carried out under reductant concentrations of 2 g/L, 4 g/L, 6 g/L, and 8 g/L at an initial NO_3^--N concentration of 50 mg/L at 30 °C. The denitrification performance of nZVI/(ABS + EC) - (0.6–0.8) was similar to that of nZVI, as has been previously reported [47] (Figure 10). The denitrification rate was observed to increase as the reductant concentration increased. However, the test results show that the denitrification rate rapidly increased when the reductant concentration increased from 2 g/L to 4 g/L, and then the rate increased slowly at reductant concentrations between 4 g/L and 8 g/L. These results demonstrate that the concentration of nZVI/(ABS + EC) significantly influenced the denitrification rate.

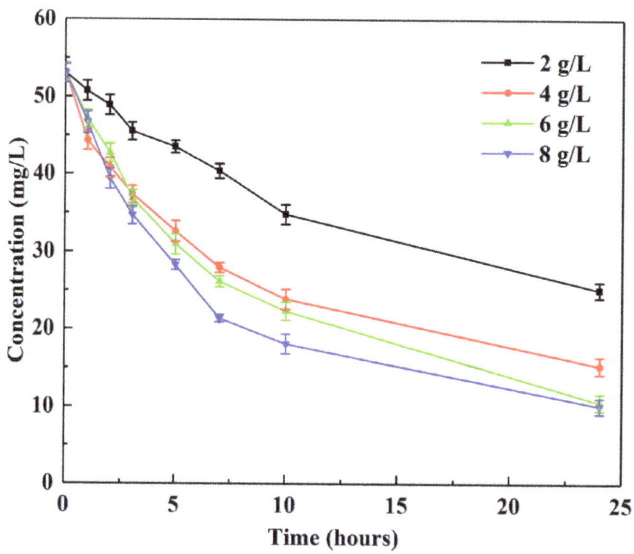

Figure 10. The denitrification performance of nZVI/(ABS + EC) - (0.6–0.8) under different reductant concentrations.

3.9. *The Denitrification Mechanism of nZVI/(ABS + EC)*

Previous studies have revealed that the deposition of passive iron oxides generated on the nZVI surface leads to the conglutination of nZVI particles, which would hinder the transport of electrons and transfer of mass for denitrification. Therefore, the above deposition on the nZVI particle surface can aggravate the aggregation and reactivity loss (Figure 11). For nZVI/(ABS + EC), embedding nZVI particles on a hole wall of the ABS matrix restricts the movement, thereby preventing the aggregation of nZVI particles. Moreover, the deposition of passive iron oxides generated on the ABS matrix surface, which can be confirmed in Figure 3d, can alleviate the reactivity loss. In addition, the regulation of the hole size and numbers can achieve the quantitative control of the sustainable denitrification performance, which is verified by the above result. Moreover, compared to regulating the denitrification rate of nZVI through functional group grafting modification of encapsulated shell materials [24], this study has the advantages of simple process and better controllability. That is, the encapsulation of nZVI in the ABS matrix can achieve the immobilization of nZVI particles and control the contact between the contaminant and nZVI particles, further accomplishing the anti-aggregation and control of sustainable denitrification performance of nZVI.

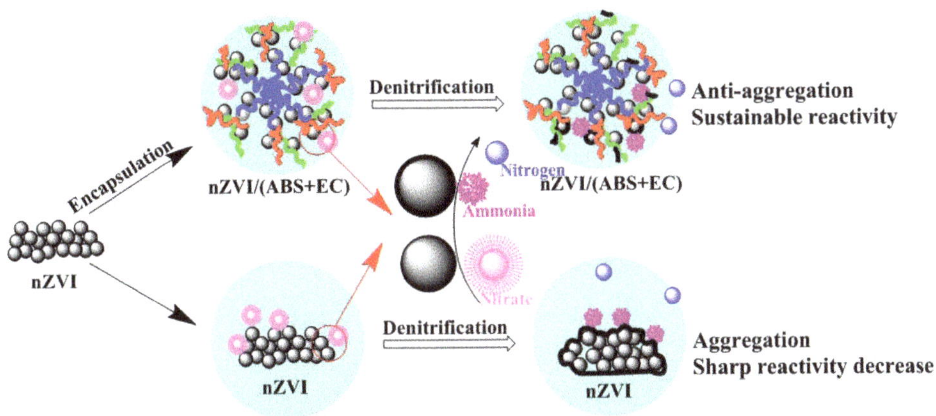

Figure 11. Denitrification mechanism of nZVI/(ABS + EC).

4. Conclusions

In this study, nZVI was successfully encapsulated in ABS for sustainable denitrification performance and anti-aggregation. The experiments demonstrated that the modification of pores within the ABS matrix using EC played a critical role in maintaining continuous nZVI reactivity. Quantitative analysis revealed a regulated denitrification rate, with apparent rate constants (k_{obs}) for denitrification performance varying from 0.0423 to 0.2036 min^{-1} across different EC and nZVI dosages. Therefore, encapsulation solves the issues of aggregation and poor reactivity persistence with the use of nZVI as the reactive medium for PRB.

Further, the denitrification rate was also affected by the pH, initial NO_3^--N concentration, and nZVI/(ABS + EC) concentration. A particularly interesting observation was the dependence of the observed rate constant (k_{obs}) on the initial NO_3^--N concentration, which was meticulously analyzed through first-order kinetic modeling. The study revealed that at varying initial NO_3^--N concentrations, the denitrification mechanism of nZVI/(ABS + EC) exhibited complex behavior, underscoring the intertwined roles of adsorption and reduction processes in NO_3^--N removal. However, the denitrification mechanism of nZVI must be thoroughly and quantitatively examined to precisely explain the observed complicated denitrification performance of nZVI/(ABS + EC) under the experimental conditions in this study. Moreover, for a comprehensive understanding of the economic efficiency of nZVI/(ABS + EC) for denitrification, future work will explore a detailed cost–benefit analysis, considering factors such as long-term operation, maintenance, and potential scalability of the nZVI/(ABS + EC) system for widespread environmental remediation applications.

Author Contributions: Conceptualization, F.M.; Methodology, M.L.; Formal analysis, Q.Z.; Investigation, Y.Y.; Data curation, C.Y.; Writing—original draft, F.M.; Supervision, B.Q.; Project administration, F.M.; Funding acquisition, F.M. All authors have read and agreed to the published version of the manuscript.

Funding: This research was funded by National Natural Science Foundation of China [grant number 42207083].

Data Availability Statement: Data are contained within the article.

Conflicts of Interest: All authors except Miao Li was employed by the company "SINOPEC Research Institute of Petroleum Processing Co., Ltd.". Miao Li declares that the research was conducted in the absence of any commercial or financial relationships that could be construed as a potential conflict of interest.

References

1. Bahrami, F.; Yu, X.; Zou, Y.; Sun, Y.; Sun, G. Impregnated calcium-alginate beads as floating reactors for the remediation of nitrate-contaminated groundwater. *Chem. Eng. J.* **2020**, *382*, 122774. [CrossRef]
2. Wakida, F.T.; Lerner, D.N. Non-agricultural sources of groundwater nitrate: A review and case study. *Water Res.* **2005**, *39*, 3–16. [CrossRef] [PubMed]
3. Hamad, J.R.J.; Yaacob, W.Z.; Omran, A. Quality Assessment of Groundwater Resources in the City of Al-Marj, Libya. *Processes* **2021**, *9*, 154. [CrossRef]
4. Suthar, S.; Bishnoi, P.; Singh, S.; Mutiyar, P.K.; Nema, A.K.; Patil, N.S. Nitrate contamination in groundwater of some rural areas of Rajasthan, India. *J. Hazard. Mater.* **2009**, *171*, 189–199. [CrossRef] [PubMed]
5. Rivett, M.O.; Buss, S.R.; Morgan, P.; Smith, J.W.N.; Bemment, C.D. Nitrate attenuation in groundwater: A review of biogeochemical controlling processes. *Water Res.* **2008**, *42*, 4215–4232. [CrossRef] [PubMed]
6. Huang, C.; Wang, H.; Chiu, P. Nitrate reduction by metallic iron. *Water Res.* **1998**, *32*, 2257–2264. [CrossRef]
7. Ruhl, A.S.; Jekel, M. Impacts of Fe(0) grain sizes and grain size distributions in permeable reactive barriers. *Chem. Eng. J.* **2012**, *213*, 245–250. [CrossRef]
8. Fu, F.; Dionysiou, D.D.; Liu, H. The use of zero-valent iron for groundwater remediation and wastewater treatment: A review. *J. Hazard. Mater.* **2014**, *267*, 194–205. [CrossRef] [PubMed]
9. Zarime, N.A.; Solemon, B.; Wan Yaacob, W.Z.; Jamil, H.; Che Omar, R.; Rafek, A.G.; Roslan, R. Adsorption of Methylene Blue by Bentonite Supported Nano Zero Valent Iron (B-nZVI). *Processes* **2023**, *11*, 788. [CrossRef]
10. Yang, G.C.C.; Lee, H. Chemical reduction of nitrate by nanosized iron: Kinetics and pathways. *Water Res.* **2005**, *39*, 884–894. [CrossRef]
11. Abdelfatah, A.M.; Fawzy, M.; Eltaweil, A.S.; El-Khouly, M.E. Green Synthesis of Nano-Zero-Valent Iron Using Ricinus Communis Seeds Extract: Characterization and Application in the Treatment of Methylene Blue-Polluted Water. *ACS Omega* **2021**, *6*, 25397–25411. [CrossRef] [PubMed]
12. Huang, T.; Liu, L.; Zhang, S.; Xu, J. Evaluation of electrokinetics coupled with a reactive barrier of activated carbon loaded with a nanoscale zero-valent iron for selenite removal from contaminated soils. *J. Hazard. Mater.* **2019**, *368*, 104–114. [CrossRef] [PubMed]
13. Maamoun, I.; Eljamal, O.; Khalil, A.M.E.; Sugihara, Y.; Matsunaga, N. Phosphate Removal Through Nano-Zero-Valent Iron Permeable Reactive Barrier; Column Experiment and Reactive Solute Transport Modeling. *Transp. Porous Med.* **2018**, *125*, 395–412. [CrossRef]
14. Tosco, T.; Petrangeli Papini, M.; Cruz Viggi, C.; Sethi, R. Nanoscale zerovalent iron particles for groundwater remediation: A review. *J. Clean. Prod.* **2014**, *77*, 10–21. [CrossRef]
15. Shao, Y.; Gao, Y.; Yue, Q.; Kong, W.; Gao, B.; Wang, W.; Jiang, W. Degradation of chlortetracycline with simultaneous removal of copper (II) from aqueous solution using wheat straw-supported nanoscale zero-valent iron. *Chem. Eng. J.* **2020**, *379*, 122384. [CrossRef]
16. Eltaweil, A.S.; El-Tawil, A.M.; Abd El-Monaem, E.M.; El-Subruiti, G.M. Zero Valent Iron Nanoparticle-Loaded Nanobentonite Intercalated Carboxymethyl Chitosan for Efficient Removal of Both Anionic and Cationic Dyes. *ACS Omega* **2021**, *6*, 6348–6360. [CrossRef] [PubMed]
17. Ruiz-Torres, C.A.; Araujo-Martínez, R.F.; Martínez-Castañón, G.A.; Morales-Sánchez, J.E.; Guajardo-Pacheco, J.M.; González-Hernández, J.; Lee, T.; Shin, H.; Hwang, Y.; Ruiz, F. Preparation of air stable nanoscale zero valent iron functionalized by ethylene glycol without inert condition. *Chem. Eng. J.* **2018**, *336*, 112–122. [CrossRef]
18. Zhang, J.; Zhang, G.; Zheng, K.; Cai, D.; Wu, Z. Reduction of Cr (VI) by urea-dispersed nanoscale zero-valent iron. *J. Nanosci. Nanotechnol.* **2015**, *15*, 6103–6107. [CrossRef] [PubMed]
19. Zhang, J.; Zhang, G.; Wang, M.; Zheng, K.; Cai, D.; Wu, Z. Reduction of aqueous Cr VI using nanoscale zero-valent iron dispersed by high energy electron beam irradiation. *Nanoscale* **2013**, *5*, 9917–9923. [CrossRef] [PubMed]
20. Fan, D.; O'Carroll, D.M.; Elliott, D.W.; Xiong, Z.; Tratnyek, P.G.; Johnson, R.L.; Garcia, A.N. Selectivity of nano zerovalent iron in in situ chemical reduction: Challenges and improvements. *Remediat. J.* **2016**, *26*, 27–40. [CrossRef]
21. O'Connor, D.; Hou, D.; Ok, Y.S.; Song, Y.; Sarmah, A.; Li, X.; Tack, F.M. Sustainable in situ remediation of recalcitrant organic pollutants in groundwater with controlled release materials: A review. *J. Control. Release* **2018**, *283*, 200–213. [CrossRef] [PubMed]
22. Matlochová, A.; Plachá, D.; Rapantová, N. The application of nanoscale materials in groundwater remediation. *Pol. J. Environ. Stud.* **2013**, *22*, 1401–1410.
23. Karimi, B.; Samadi, S. Nitrate removal from aqueous solution: Using zero-valent iron (Fe degrees) fixed on Ca-alginate bead. *Desalin. Water Treat.* **2019**, *143*, 235–239. [CrossRef]
24. Krajangpan, S.; Bermudez, J.J.E.; Bezbaruah, A.N.; Chisholm, B.J.; Khan, E. Nitrate removal by entrapped zero-valent iron nanoparticles in calcium alginate. *Water Sci. Technol.* **2008**, *58*, 2215–2222. [CrossRef] [PubMed]
25. Jiang, Z.; Lv, L.; Zhang, W.; Du, Q.; Pan, B.; Yang, L.; Zhang, Q. Nitrate reduction using nanosized zero-valent iron supported by polystyrene resins: Role of surface functional groups. *Water Res.* **2011**, *45*, 2191–2198. [CrossRef] [PubMed]
26. Wang, X.; Zhang, B.; Ma, J.; Ning, P. Chitosan Modifying Nanoscale Zero Valent Iron for Tetracycline Removal from Aqueous Solutions: Proposed Pathway. *Environ. Eng. Sci.* **2019**, *36*, 273–282. [CrossRef]

27. Choi, H.; Al-Abed, S.R.; Agarwal, S. Effects of aging and oxidation of palladized iron embedded in activated carbon on the dechlorination of 2-chlorobiphenyl. *Environ. Sci. Technol.* **2009**, *43*, 4137–4142. [CrossRef] [PubMed]
28. Wang, X.; Yi, D.; Ma, J. Novel synthesis of carbon spheres supported nanoscale zero-valent iron for removal of metronidazole. *Appl. Surf. Sci.* **2016**, *390*, 50–59. [CrossRef]
29. Su, C.; Puls, R.W.; Krug, T.A.; Watling, M.T.; O'Hara, S.K.; Quinn, J.W.; Ruiz, N.E. A two and half-year-performance evaluation of a field test on treatment of source zone tetrachloroethene and its chlorinated daughter products using emulsified zero valent iron nanoparticles. *Water Res.* **2012**, *46*, 5071–5084. [CrossRef]
30. Ferrari, P.F.; Trucillo, P.; De Negri Atanasio, G.; Bufalini, C.; Campardelli, R.; Perego, P.; Palombo, D.; Reverchon, E. Operating Parameters Optimization for the Production of Liposomes Loaded with Antibodies Using a Supercritical Fluid-Assisted Process. *Processes* **2023**, *11*, 663. [CrossRef]
31. Mora-Huertas, C.E.; Fessi, H.; Elaissari, A. Polymer-based nanocapsules for drug delivery. *Int. J. Pharmaceut.* **2010**, *385*, 113–142. [CrossRef] [PubMed]
32. Guan, H.; Chi, D.; Yu, J.; Li, H. Encapsulated ecdysone by internal gelation of alginate microspheres for controlling its release and photostability. *Chem. Eng. J.* **2011**, *168*, 94–101. [CrossRef]
33. Gonzalez Gomez, A.; Syed, S.; Marshall, K.; Hosseinidoust, Z. Liposomal Nanovesicles for Efficient Encapsulation of Staphylococcal Antibiotics. *ACS Omega* **2019**, *4*, 10866–10876. [CrossRef]
34. Kavetsou, E.; Koutsoukos, S.; Daferera, D.; Polissiou, M.G.; Karagiannis, D.; Perdikis, D.; Detsi, A. Encapsulation of Mentha pulegium essential oil in yeast cell microcarriers: An approach to environmentally friendly pesticides. *J. Agric. Food Chem.* **2019**, *67*, 4746–4753. [CrossRef]
35. Lin, G.; Chen, X.; Zhou, H.; Zhou, X.; Xu, H.; Chen, H. Elaboration of a feather keratin/carboxymethyl cellulose complex exhibiting pH sensitivity for sustained pesticide release. *J. Appl. Polym. Sci.* **2019**, *136*, 47160. [CrossRef]
36. Slattery, M.; Harper, B.; Harper, S. Pesticide Encapsulation at the Nanoscale Drives Changes to the Hydrophobic Partitioning and Toxicity of an Active Ingredient. *Nanomaterials* **2019**, *9*, 81. [CrossRef]
37. Maes, C.; Menot, B.; Hayouni, S.; Martinez, A.; Fauconnier, M.; Bouquillon, S. Preparation of New Glycerol-Based Dendrimers and Studies on Their Behavior toward Essential Oil Encapsulation. *ACS Omega* **2022**, *7*, 10277–10291. [CrossRef] [PubMed]
38. Fugolin, A.P.P.; Ferracane, J.L.; Pfeifer, C.S. Fatigue-Crack Propagation Behavior in Microcapsule-Containing Self-Healing Polymeric Networks. *Mater. Des.* **2022**, *223*, 111142. [CrossRef]
39. Gharsallaoui, A.; Roudaut, G.; Chambin, O.; Voilley, A.; Saurel, R. Applications of spray-drying in microencapsulation of food ingredients: An overview. *Food Res. Int.* **2007**, *40*, 1107–1121. [CrossRef]
40. Augustin, M.A.; Hemar, Y. Nano-and micro-structured assemblies for encapsulation of food ingredients. *Chem. Soc. Rev.* **2009**, *38*, 902–912. [CrossRef] [PubMed]
41. Rubio-Rodríguez, N.; Beltrán, S.; Jaime, I.; Sara, M.; Sanz, M.T.; Carballido, J.R. Production of omega-3 polyunsaturated fatty acid concentrates: A review. *Innov. Food Sci. Emerg.* **2010**, *11*, 1–12. [CrossRef]
42. Paulo, F.; Santos, L. Design of experiments for microencapsulation applications: A review. *Mater. Sci. Eng. C* **2017**, *77*, 1327–1340. [CrossRef] [PubMed]
43. Meng, F.; Wang, S.; Liu, H.; Xu, X.; Ma, H. Microencapsulation of oxalic acid (OA) via coacervation induced by polydimethylsiloxane (PDMS) for the sustained release performance. *Mater. Des.* **2017**, *116*, 31–41. [CrossRef]
44. Meng, F.; Li, M.; Wang, S.; Liu, X.; Gao, W.; Ma, Z.; Kong, C.; Ma, X.; Li, J. Encapsulation of potassium persulfate with ABS via coacervation for delaying the viscosity loss of fracturing fluid. *J. Appl. Polym. Sci.* **2019**, *136*, 47734. [CrossRef]
45. Xu, J.; Hao, Z.; Xie, C.; Lv, X.; Yang, Y.; Xu, X. Promotion effect of Fe^{2+} and Fe_3O_4 on nitrate reduction using zero-valent iron. *Desalination* **2012**, *284*, 9–13. [CrossRef]
46. Wang, W.; Jin, Z.; Li, T.; Zhang, H.; Gao, S. Preparation of spherical iron nanoclusters in ethanol–water solution for nitrate removal. *Chemosphere* **2006**, *65*, 1396–1404. [CrossRef] [PubMed]
47. Zhang, J.; Hao, Z.; Zhang, Z.; Yang, Y.; Xu, X. Kinetics of nitrate reductive denitrification by nanoscale zero-valent iron. *Process Saf. Environ.* **2010**, *88*, 439–445. [CrossRef]
48. Suzuki, T.; Moribe, M.; Oyama, Y.; Niinae, M. Mechanism of nitrate reduction by zero-valent iron: Equilibrium and kinetics studies. *Chem. Eng. J.* **2012**, *183*, 271–277. [CrossRef]
49. Noubactep, C. Comments on 'Mechanism study of nitrate reduction by nano zero valent iron' by Hwang et al. [J. Hazard. Mater. (2010). https://doi.org/10.1016/j.jhazmat.2010.10.078]. *J. Hazard. Mater.* **2011**, *186*, 946–947. [CrossRef] [PubMed]

MDPI AG
Grosspeteranlage 5
4052 Basel
Switzerland
Tel.: +41 61 683 77 34

Processes Editorial Office
E-mail: processes@mdpi.com
www.mdpi.com/journal/processes

www.ingramcontent.com/pod-product-compliance
Lightning Source LLC
Chambersburg PA
CBHW062050220926
51850CB00010B/2716
* 9 7 8 3 7 2 5 8 4 7 8 3 9 *